Guido Wilke und Jürgen Ortmeier

Coatings for Plastics

Wilke/Ortmeier: Coatings for Plastics

ISBN: 978-3-86630-859-6

Cover: Karl Wörwag Lack- und Farbenfabrik GmbH & Co. KG

Guido Wilke and Jürgen Ortmeier
Coatings for Plastics
Hanover: Vincentz Network, 2012
European Coatings Tech Files
ISBN 3-86630-859-0
ISBN 978-3-86630-859-6

Please ask for our book catalogue
Vincentz Network, Plathnerstr. 4c, 30175 Hanover, Germany
T (202) 684-6630, F (202) 380-9129
E-mail: books@american-coatings.com, www.american-coatings.com

Layout: Danielsen Mediendesign, Hanover, Germany

ISBN 3-86630-859-0
ISBN 978-3-86630-859-6

European Coatings Tech Files

Guido Wilke und Jürgen Ortmeier

Coatings for Plastics

Wilke/Ortmeier: Coatings for Plastics

ISBN: 978-3-86630-859-6

Foreword

Today's innovations in the coatings industry are a result from close collaboration between coating material suppliers and its partners. In general, concerning modern industrial coatings, it is not sufficient for paint supplier just to head for any innovative material with specific technical benefits, but to become involved early in a joint development of the objects to be coated. The development of automotive exterior parts, where the ratio of plastic substrates has increased continuously over the last decades, might serve as a good example for this trend. In the 1970s plastic assembly parts of a vehicle could easily be recognized as dark grey coloured articles mounted after the coating process, whereas today form integration and body colour is the rule.

At the same time plastic substrates have found increasing acceptance for household appliances and objects of daily use. For decorative and functional reasons many of these surfaces are coated. Since its beginnings this development underwent an enlargement of the scientific and technological basis on different levels and an ongoing exchange of know-how between those levels.

This book addresses the whole circle of participants in the technology of manufacturing coated plastic parts. It may impart developers of plastic coatings, the technical service staff members of coating and raw material suppliers as well as coating processors with the know-how, which is needed for a sustainable successful manufacturing of coated plastic goods. In fact, a firm understanding of its elements and their complex dependencies is necessary in order to benefit from the advantages of coating plastics and to avoid cost intensive failures.

The approach of this book is to present the technology of coating plastics in a broad and integrated mode, treating both the single influence fields and their connections. One focus is set on the plastic substrate itself, as it is the origin of many characteristic issues. Substrate topics are then connected with coating relevant plastic processing aspects and surface pretreatment methods. The chapter of coating materials presents a broad portfolio of modern formulation and layer concepts, their main characteristics with respect to end users requirements. Paint application, process and equipment technology is treated in a compact way, as many aspects coming from this direction for coating moulded plastic components are similar to those of coating general industrial goods. Coatings on plastics have to be tested, so the main characteristics of those polymer layer composites will also be treated. The most frequent coating defects, their detection, possible causes and prevention are presented in the final chapter.

This publication is the English translation of the original publication written in German. We gladly thank Mr. Steve Sadlak for his support in providing assistance in translating its contents to English text.

Esslingen and Stuttgart, in October 2011
Dr. Guido Wilke and Jürgen Ortmeier

Wilke/Ortmeier: Coatings for Plastics

ISBN: 978-3-86630-859-6

Exclusion of liability

Please note, the selected formulas are representative examples for the respective applications as outlined, and are derived from raw material supplier's recommendations and available published patent literature. If there is no general state of the art established to date, no formula is specified. Also note, it is never possible to use the formulas as suggested without further development and testing.

Any liability for technical advice is expressively excluded.

Restrictions by patents and other intellectual property rights however cannot be excluded. Trademarks (e.g. ® or ™) are not explicitly referenced. Trademarks and supply sources are subject to change.

Contents

Wilke/Ortmeier: Coatings for Plastics

ISBN: 978-3-86630-859-6

1 Introduction

1.1 History and Market trends

Starting with applications like TV-housings, sports goods, automotive rear spoilers and buttons, in the 60s and 70s of the 20th century, the coating of plastics, after getting over some early design challenges, has developed into an innovative technology. In particular, the finishing and decoration of plastic surfaces has become a large industry. A guiding effect to today's industry came from the evolution of automotive manufacturing. Almost every visible plastic part of an automobile today is coated. Today's state of the art plastic coatings resulted from the joined work of diverse segments of production technology leading to specific innovations and adaptations. Important milestones of development were the initial design of performance plastic resins, followed by advances in injection moulding processes and tool design, coupled with the development of coating chemistries which meet the OEM demands for which the part is painted. In the beginning solvent-borne systems were dominating the marketplace, and were gradually replaced by low-emission coatings due to environmental restrictions and higher levels of performance. For example the coating of large automotive exterior parts with water-borne materials of colour and effect identical to the car body today is state of the art. But the achieved progress in plastic coatings is not limited to the automobile (Table 1.1).

Table 1.1: Application areas and examples of plastic coatings

Application area	Examples
Automotive	Bumper covers, body panels, trim, headlamp lenses, mirror housings, instrument panel covers, air bag covers
Commercial electronics	Covers and housings for cell phones, PSA's, computer housings, TV, MP3-players
Household appliances	Covers for vacuum cleaners, washing machines
Medical technology	Glasses, catheters, infusion bags
Information technology	CDs, smart cards
Packaging	Food, drinks, snacks, cleaning agents
Building	flooring, window frames, doors, furniture
Sporting goods	Skis, hulls, eyewear, racing cars, helmets

Wilke/Ortmeier: Coatings for Plastics

ISBN: 978-3-86630-859-6

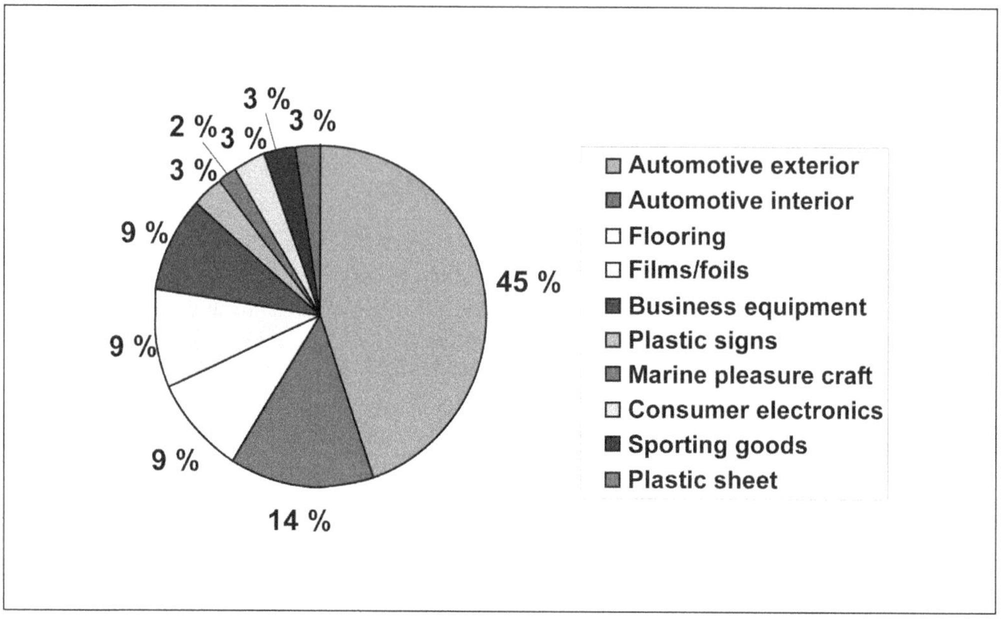

Figure 1.1: Market of plastic coatings in North America *Source: The ChemQuestGroup* [1]

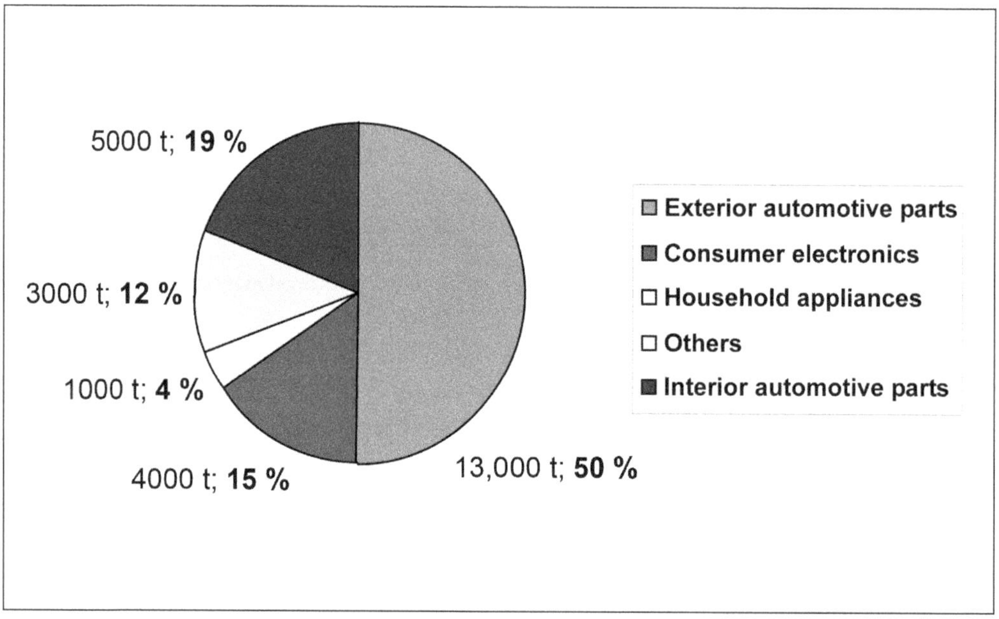

Figure 1.2: Market of plastic coatings in Germany

Source: Yearly report 2005/2006; Federation of the German coatings industry (VdL), from [2]

Reliable and detailed data about the market of plastic coating systems are rare and hardly published. Figures 1.1 and 1.2 present the distribution of business areas for plastic coatings in North America and Germany.

1.2 Purpose of coatings on plastics

The purposes of coatings on plastics can be summarized into three groups:

- protection,
- decoration and
- special functions.

1.2.1 Protection

Most plastics are not resistant to the elements of weather. The appearance of plastic surfaces typically gets worse over time due to complex interaction of climatic factors like UV radiation, oxygen, pollutants as well as variations of temperature and air humidity. Coatings equipped with UV-absorbers and radical scavengers' help plastics overcome some level of degradation but in order to be weather resistant require the use of expensive additive packages.

Many thermoplastic plastics only have a limited resistance to abrasion, scratch, solvent and chemical attack. Printed and metalized plastics are often protected against abrasion and corrosion by a thin layer of coatings, that thanks to their polymer composition and network character provide those objects with the necessary durability. Full protection coatings must adhere well not only initially, but also after aging, weathering, mechanical stress and contact to diverse liquid solutions.

As plastics materials consist of very different mechanical profiles, one of the conditions for coating plastics is, that the coating should not be more rigid than the substrate, but its flexibility should be at best adapted to that of the plastic substrate. In general this requisite today can be realized quite well, thanks to a large selection of suitable organic binders. The importance of protective functions is reflected in some of today's current trends. One of the priorities in the development of modern coatings is the concept of hybrid materials, consisting of organic polymers that are modified with inorganic components or nano-scaled extenders. Hybrid-materials containing suitable organo-polymers and ceramic substances allow almost any combination of typical benefits of both classes of materials, resulting in flexible but also durable coatings for plastic surfaces of high decorative value. Since UV-cured scratch resistant coatings on plastics like polycarbonate and "Plexiglas" are considered somewhat state of the art, other options may include radiation curing to avoid the use of solvents and do offer additional benefits including increased curing speed and the processing equipment footprint is minimal.

1.2.2 Decoration

Plastic parts are coated today with impressive variability in terms of colour, gloss and effect. Examples are the colourful design of mobile phones as well as automobiles, which to the layman do not show any big colour difference between the car body and supplier parts. But although this colour harmony seems so natural, it has to be achieved again and again for every new process, model and colour. While pigmented coatings conceal the original colour of the plastic part, the gloss and chemical resistance of mass coloured plastic surfaces are improved by clear coats.

The fact that moulded plastic parts can be coated successfully at all is the result of many years of research to raise the surface quality to a coat able state. Today's plastic processing does not always yield class A surfaces. The decorative painting of problematic plastic substrates at first affords an elimination of visible surface defects.

Flow patterns and seams of thermoplastic injection moulded parts telegraph through a decorative coating layer. Shrink-holes and other defects of long fibre reinforced duroplastic polymers also have to be levelled properly. Components made with reinforced thermoplastic polymers like polyamide often show an increased roughness due to short fibres standing out from the plastic surface. The levelling of the surface can in principle be achieved with special primers and surfacers.

1.2.3 Special properties

Besides protective and decorative tasks, there are other functions that plastic coatings may execute, in order to fulfil more specific demands. Some examples are listed below:

- Formation of diffusion barriers on packaging foils against the permeation of gaseous and liquid substances from the atmosphere or the packaging goods.
- The preparation of plastic surfaces for metallization.
- The creation of surfaces with a specific haptic character (automotive interior applications).
- The absorption or reflection of certain electromagnetic waves (aircrafts, military vehicles).
- The reduction of friction of squeegees for window wiper blades.

Coating technology can also be used to supply a plastic part with properties that are not an intrinsic part of their original technical profile rising from its chemical composition. Examples are coatings with electrical or magnetic conductivity, allowing adding those properties to plastics without modification to the bulk polymer.

Modern functional coatings in general can be realized on various substrates, so on plastics. For example intrinsic conductive polymers find use in antistatic coatings for plastic foils in electronic packaging [3]. The flexibility of those polymers allows the integration of electronic printed circuits in textile clothes and packaging. Hereto those polymers are printed on foils as an electrical conductive circuit [4].

1.3 Challenges of today and of the future

1.3.1 Increase of process stability

The complexity of coating plastics calls for incidence of flaws on coated parts rising from diverse spheres of influence: substrate, plastic processing, pre-treatment, paint material and paint application. Intense efforts are concentrated on reducing coating repair work in order to increase the first run ok-rate.

Substrate generated coating failures are a frequent quality issue with fibre reinforced plastics. So the material SMC *(sheet moulding compound)* tends to gas out, resulting in popping and pinholes. In the painting process one may take account for this characteristic by reducing the thermal impact as much as possible in order to minimize the evaporation of volatile plastic compounds. This may require either long curing times at low temperatures or alternative cure concepts. Of course the process ability of the latter has to be accessed profoundly, before introducing it into production.

In the field of automotive coatings the colour harmony between add on components and the car body is a topic of enduring importance. Significant progress has been achieved within the last few years by various measures, like the limited tolerance of colour quadrants, the implementation of colour measurement systems and by the adaptation of paint application techniques for plastic parts and car body. One further option is the standardization of basecoat formulations for car body and plastic parts. However, with every new colour the usefulness of the existing concepts for colour harmony is put to the test again.

Concerning the technical coating properties, the main issue of adhesion is neither solved for all plastic substrates nor is it well understood on a scientific base. To secure paint adhesion on polyolefin's is the biggest challenge, moreover as they gain more and more market share. For automotive OEM coatings, scratch resistance is also a well known demand, and meanwhile it has established itself also in the field of coating add on plastic components.

1.3.2 Reduction of emissions/low emission coatings

Many of the larger Tier One component suppliers have recently switched to waterborne coatings, including the adaptation of specialized humidity and temperature controlled processes for their application. However, the shift from conventional solventborne to low-VOC clear coats is a task still waiting to be solved for many fields of application. Besides waterborne materials, another option for clear coats is UV curing systems. Material development has left the state of basic research and development and finds evaluation underway in many market segments, with degrees of success depending on the operational area. Powder coatings are only suitable for plastics with low thermal expansion, so that parts can overcome higher cure temperatures without loss of shape. It is not expected that in major operational areas like the automotive supplier parts business, powder coatings would substitute the classical wet paints in the foreseeable future.

1.3.3 Increase of efficiency/profitability

Dispensing with coating layers is one option by which producers of automotive supplier parts reduce costs. Positive experiences have been made with the primer-less coating of bumper housings. However an important demand of this approach is a high plastic surface quality. So in cases of difficult substrates with lower surface quality, for example with some highly filled fibre reinforced plastics, one may not be successful without the use of a primer. If the substrate quality demands are fulfilled, as in the case of some coating processes for bumper-housings, this concept may work. Here a two layer paint body is realized with success. It consists of a (waterborne) basecoat and clear coat. In other cases, the requirements of colour harmony and intrinsic colour of the substrate, in combination with a low hiding power of the basecoat, might afford the use of a coloured primer.

Dispensing with coating layers in its most extreme way has been practiced by decorating mass coloured plastic parts utilizing just a clear coat. By this method straight shades can be realized in an elegant way, as it is done with body panels, like the "Smart" car.

An extension of clear-coated mass-coloured plastics on other car models would afford the realization of effect colours. But the mass colouring of plastics in this special case is confronted with the high demand of colour and effect harmony on the car. As the pigmentation of the plastic bulk naturally does not result in an orientation of effect pigments similar to that of a paint spray application, up to now it has not prevailed as a successful concept for coating automotive assembly parts.

In the search of alternative coating concepts a completely different direction is followed by foil technology. In connection with the possibility to combine several car body parts to one module, foil technology promises to be particularly cost efficient by reducing the vertical range of manufacture. Meanwhile there are numerous foil concepts, which will be further addressed in Chapter 4, and have found limited introduction into the market. Current applications are limited to extensive body parts of moderately complex geometry. One typical and appropriate application for foil concepts is the car roof that can be partitioned in rigid, movable, well hiding and transparent modules. One might hope and expect that foil technology will master its challenges and continue to gain in popularity.

1.3.4 Design

One of the major trends currently observed in the market is individualization. The wish for colour and effect is considered as the driving force for an ongoing market growth for plastic coatings, especially for vehicles. Also in business areas like commercial electronics, represented by objects such as mobile phones, notebooks, etc. which are styling trends are influenced by individual colour design. Today these aesthetic needs can be realized at best by advanced coating technologies. This is particularly evident in coatings with interference and pearl effect, which develop their optical performance, such as colour flop effects, only by the interaction of the coating material and the application process.

One area, in which the trend for individualization can be developed in a strong way, is automotive interior coating. The great variety of customers needs can be at best be satisfied by many materials. The colour of plastic, leather and textile materials have to harmonize with each other. In this context the coating of automotive interior parts is necessary, because coating allows the most flexible differentiation of the optical design. A large variety of colours in the automotive interior can only be realized with plastic coatings today, but innovative foil concepts like ***insert film moulding*** (IFM) have already found its place in the market. Three dimensional shaped plastic parts, being illuminated by the application of an alternate voltage, can be manufactured by back injection moulding of a transparent foil that is then coated with a thin electroluminescent layer [5].

In general one predicts a great future for switchable coating systems. It may also be assumed that there will be further examples for the innovative connection of switchable coating systems with plastic processing, allowing expanding marketplace acceptance when design and function of industrial products are desired and/or needed.

1.4 References

[1] The ChemQuestGroup, Inc., www.radtech.org/An Overview of the North American Plastic Coatings.pdf, 2003

[2] U. Hoffmann, Journal für Oberflächentechnik (JOT) 46, 11 (2006), 56

[3] S. Kirchmeyer, L. Brassat, Kunststoffe 95, 10 (2005), 202

[4] H. Rost, Kunststoffe 95, 10 (2005), 209

[5] E. Foltin, G. Wießmeier, Journal für Oberflächentechnik (JOT) 3 (2004), 34

2 Plastics in coating technology

2.1 Survey of plastic materials

2.1.1 Classification by technical criteria

According to DIN 7724, plastics can be classified into three groups: thermoplastic , elastomeric and thermosetting (duroplastic) materials.

The basic characteristics of these three groups of materials as well as their international abbreviations according to DIN EN ISO 1043-1, set in brackets, shall be presented here briefly.

Thermoplastic polymers are not cross-linked, more or less soluble and can be softened or molten reversibly. Materials belonging to this largest group of polymers are the so-called standard plastics like polyethylene (PE), polypropylene (PP), polystyrene (PS) and polyvinylchloride (PVC). Also many of the so-called technical plastics belong to this group, examples are polyesters like polyethylene terephthalate (PET) and polybutylene terephthalate (PBT), polyamides (PA), blends of polystyrene- (ASA, ABS, SB, SBS), polycarbonate (PC) and its blends (PC/PBT, PC/ABS), and polymethyl methacrylate (PMMA).

Elastomers are widely cross-linked polymers, showing a rubber-like at ambient temperature. They are generally not soluble in organic solvents, but swellable, and can be softened by heating, but then they don't have a plastic character and therefore cannot be moulded easily. Examples are natural or synthetic rubber (NR, BR), silicon rubber (SI), some polyurethanes (PUR) and fluorine rubber (FKM).

Thermosetting (duroplastic) polymers are densely cross-linked materials that are hard and brittle at ambient temperature. They are not soluble or swellable in most organic solvents, and upon heating they soften only slightly. Therefore, thermosets exposed to elevated temperatures are not easily processed and show some level of decomposition. Examples for thermosetting polymers are cross-linked epoxy- (EP), unsaturated polyester- (UP), phenol- (PF), melamine- (MF) and urea resins (UF).

2.1.2 Economical and coating related significance

Over the past 50 years, world production of plastics has increased on average 10 % annually, and in 2008 reached 245 Million tons [1].

Wilke/Ortmeier: Coatings for Plastics

ISBN: 978-3-86630-859-6

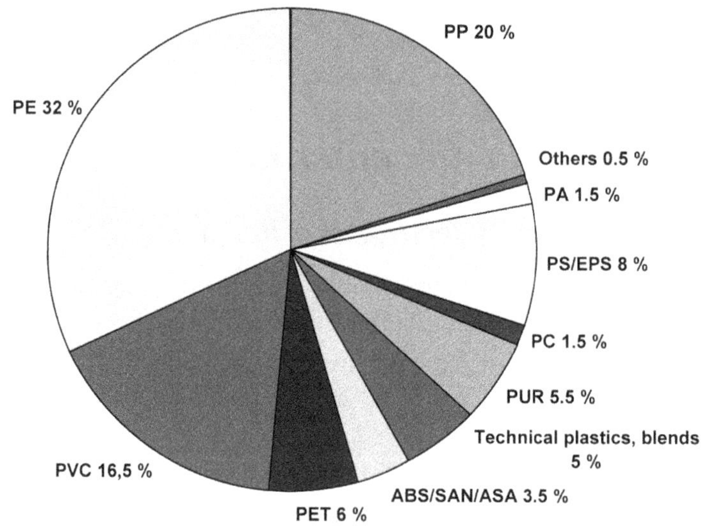

Figure 2.1: Worldwide production of plastics in 2004 *Source: Verband Kunststofferzeugende Industrie, from [1]*

Figure 2.1 presents the share of market segments on a worldwide basis in 2004 and Table 2.1 shows those plastic substrates being most significant for coatings, and their fields of application.

Roughly half of the worldwide consumption of plastics accounts for **polyolefins**. This group is represented mainly by **polyethylene** (PE), which has the biggest share and by **polypropylene** (PP), showing a strong increase in market share (6 % per year on a worldwide basis, 12 % in china), resulting from its use for high value and cheap mass articles. It is expected that polyolefins will dominate the market even more in the future, because of their yet not fully explored high tech-potential, the ongoing price pressure on industrial products, and due to the constraints of recycling. About 70 % of all automotive plastic materials are based on polypropylene [2].

Pure and coated polyolefins are important materials for the packaging industry. For example, foils of polyethylene and polypropylene are coated with barrier layers and hollow pieces

Table 2.1: Examples for plastics in coatings technology

Business area	Plastics (examples)
Packaging	PE, PP, PET
Automotive	TPO, SMC, PA6-GF, PC+ABS, PUR-RRIM, PC+PBT, PPO+PA, PA+ABS
Aircraft	CFK (EP-CF)
Commercial electronics	PS, styrene blends, PC, PP
Medicine	PC, PP
Information technology	PC, PVC
Building trade	PVC, PS, EPS, PUR
Sports	GFK, CFK, PC, ABS

are printed for decorative purposes. Flexible PP-blends like TPO have a strong position in the car industry). The share of TPO in plastics used for automobiles is about 20 %. Examples of items are deep drawn foils for dashboards and injection moulded bumper fascia. Efficiency, recyclability and safety issues (passive pedestrians safety) cause an increasing trend towards replacing other common plastics with polyolefins.

Another standard plastic, that is processed to foils and sheet flooring, and frequently coated afterwards is **polyvinyl chloride** (PVC). The group of **polystyrenes** (PS) and foamed, expanded polystyrenes (EPS), since years ranking forth in worldwide production, is mainly applied in the packaging and building industry. Television cabinets of polystyrene were among the first plastic objects that were coated industrially. Other fields of applications for painted polystyrene are hi-fi systems and computer housings. As elastified polystyrene-type plastics like copolymers with butadiene (ABS) and acrylester (ASA) suit often better, today they have more significance as substrate for coatings. They have claimed a firm position in the field of consumer electronics and as a blend with polycarbonate (PC+ABS, PC+ASA) in automobile parts. Examples for coated parts from those blends are exterior mirror housings, licence plates and consoles. A reinforced ABS/PA-blend is in use for online-coatable fenders.

Thermoplasts like polycarbonate (PC), polybutylene- and polyethylene terephthalate (PBT resp. PET) and polyamide (PA) are referred to as technical plastics. With a production share of around 9 % of the world market they rather belong to the small volume plastics today, for which on the other hand the highest production growth rates of 6 to 8.5 % are forecasted. Technical plastics are popular substrates in the coatings technology. Polycarbonate is applied as material for discs and sight panes, where they have to be equipped with a thin coating layer protecting against UV light and scratches. An important application of PBT is as a blend with PC in car bodies, for example for air-inlet grilles. PET is used for coated multilayer films for food packaging (snacks), coated polyamide (PA) for example in household appliances like washing machines. Glass fibre reinforced polyamides are applied especially for tool housings, vehicle door handles, rear handles and wheel covers.

The blend of PA with the heat stable **polyphenylen oxide** (PPO) has a strong position for automobile fenders. In most cases, except for bumpers and fenders, large surface parts for automobiles and also for aircrafts and boats are made from duroplastic fibre reinforced polymers. Those fibre composites can be composed of cured **unsaturated polyesters**- (UP) or **epoxy resins** (EP) and glass fibres (GF) or carbon fibres (CF). One variant of glass fibre reinforced plastics that is of significance for large automotive vehicle parts is **SMC** (Sheet Moulding Compound). This material consists basically of glass fibre mat impregnated with UP-resin, equipped with fillers, thickeners, catalysts, further additives and rolled up. After cutting an appropriate sheet from the roll, it is inserted into the press form and cured thermally to the desired part.

Polyurethanes also are applied as substrates in coatings. By reaction of polyoles with isocyanates in a form, **PUR-RIM** (Reaction-Injection-Moulding) parts and reinforced types (PUR-RRIM) can be obtained. Front and side fascia of automobiles have been manufactured of this group of materials.

Nanocomposites present a group of polymer materials, which today is intensely investigated in research and development. It is the high ratio of surface and volume of nanoparticles that promises extreme physical properties, compared with those of normal scaled particles. Examples of nano-scaled materials, applied for reinforcing plastics, are sol-gel-materials, nano-scaled fillers, carbon nanotubes and nano-layer silicates [3].

2.1.3 Causes for the increasing importance of plastic

Why is it, that plastic has become an interesting material for industrial and consumer goods? The responsible reasons for this can be summarized as follows:

2.1.3.1 Economical and ecological aspects

Most plastics have a low density and are the materials of choice for weight reduction and lower fuel consumption of automotive vehicles and aircrafts. Compared to iron, which has a specific weight of 7.873 g/cm^3 polyolefins with densities of less than 1 g/cm^3 appear as true fly weights, offering a higher potential for light weight construction than even aluminium (2.7 g/cm^3). Foams show even lower density values. Reinforcements like glass fibres however effect an increase of polymer materials density. In order to use the options of weight reduction carbon fibre reinforced plastics are applied increasingly in the aircraft industry dominantly, but only to a minor extent in the automotive industry. Experiences have shown that a weight reduction of 10 % may result in savings of fuel consumption of about 7 % [4]. The use of plastics in automobiles is estimated to be about 15 % on average, with an equal split of exterior, interior, chassis and under-hood components [5].

The manufacturing of a given part affords less amount of plastic than it would be necessary in the case of steel. The production of one litre of polypropylene consumes energy of about 1.2 litres of crude oil. For the same volume of iron the demand is 8 litres crude oil, for aluminium it is 15 litres crude oil. Although crude oil is currently the only used raw material resource for the production of plastics, in principle, every source of carbon can be used, even lime or carbon dioxide [6].

The mass production of moulded parts at low costs is generally considered as the decisive economical advantage of plastics. Also semi-finished products like foils and sheets can be manufactured rather cheaply by efficient plastic processing. Many thermoplastic polymers can be processed eco-efficiently at low temperatures to complex parts that do not need any post-processing. This opens the way to the conservation of fossil resources and to low production costs. The manufacturing of high-value parts from thermosetting materials, often with reinforcement, in most cases is more expensive than the use and processing of thermoplastics. The quantity of parts to be produced either of plastic or steel also is an important parameter for the calculation of manufacturing costs.

There are far more possibilities to process plastic than just to manufacture simple moulded parts. Complete modules consisting of several parts can be manufactured in a decentralized way and in only a few steps, for what in former times a long chain of serially proceeded single operations was necessary. By the simplification of manufacturing concepts and the reduction of the vertical range of manufacture plastic has opened the door to new and more efficient ways of production.

2.1.3.2 Technology and design

The injection moulding of plastics has increased strongly the scope of possibilities for the design of industrial goods. A prime example is automotive body construction, where plastic allows a harmonious change of shape from assembly part to the car body. For many years exterior mirrors, bumpers and door handles no longer appear like screwed on parts, but have become integral components of the car body. Consequently the exchange of typical metal against plastic parts resulted in an increase of freedom of design. With an increasingly harmonious integration of assembly parts, the question of colour matching became more and more important. Bumpers in body colour were perceived as added value to the automobile and therefore they were soon specified as the design-standard. This happened regardless of any disadvantages, like a potential increased sensitivity of the coated surface towards mechanical damages and its more costly repair. However in some countries cars equipped with grey or black bumpers almost have disappeared from the market.

There are also some other issues where freedom of design has profited from the application of plastics: Antennas today are more and more often invisible, because of their integration in the car body. They can be incorporated into the trunk lid, made of SMC, which is a procedure, which is not possible with steel. One physical cause for this is the electromagnetic conductivity, which is much lower for plastics than for steel. The use of plastics as a material for the interior of vehicles allows the integration of various functions in one part and the installation of airbags in a space saving way.

The combination of plastics with other materials, as proceeded in the hybrid construction method offers further, yet not fully explored possibilities of alternative manufacturing concepts. One example for this is the roof module, made of glass and plastic, whose design and material combination would not be possible in that way in the case of metal.

Apart from design aspects, the variability of physical and technical characteristics is especially representative for plastics (tailor suited materials). Some of these characteristics are described in the following, whereby they partly present themselves in an ambivalent way. Often it depends on the kind of application and on the chances of material adaptation, whether the technical characteristics and the application of plastics are beneficial or disadvantageous.

2.2 Material characteristics

The physical and engineering characteristics of plastics are summarized very well in some monographs [7-9]. However, not all of the characteristics and materials presented in those works are relevant for the coatings technology. Therefore not every technical characteristic of every material is treated below, but mainly those, which allow a direct connection to issues of the coating of plastics. This applies for example the selection of material for the component design, mainly the coat-ability and effects of material properties on coating performance, as well as the influence of coatings on material and component properties. Table 2.2 presents material properties and their effects on properties of the coated plastic part.

Table 2.2: Comparison of material properties and effects on coating and part performance

Material issue	Effect
Mechanical behaviour	Coating adhesion Stone chip resistance Impact-performance
Thermal behaviour	Heat distort resistance (coating curing) Yellowing Dimension stability, residual stress, stress cracking
Solubility/swellablility	Coating adhesion Stress cracking Cohesion (impact-performance)
Electrical behaviour	ESTA paintability Electrostatic charging (dust)
Surface characteristics	Leveling (wavescan values, structure) Wetability (coatings adhesion)

2.2.1 Mechanical properties

Deformation characteristics (mechanical characteristic values)

Some fundamental mechanical insights can be obtained by deformation experiments. Standardized tension, bending, torsion and compression tests provide exact physical values. In general, one has to differ between the material behaviour under low stress and high stress. The latter usually result in irreversible deformation with failure of the material. Investigating the temperature dependant deformation under oscillating loads, like performed in the dynamic-mechanical analysis, is used to classify plastics according to DIN 7724 (Figure 2.2). Furthermore DMA allows the determination of elastic moduli, transition temperatures and network densities (see also Chapter 2.2.2).

On the other hand the response to linear increasing stress, as executed in the tensile test, may provide information about the failure behaviour of materials. Insights into the general mechanical character of polymer materials can be obtained by the tensile test according to ISO 527 (Figure 2.3). The popular characterization of polymers as "stiff", "brittle", "strong" or "tough" can be quantified by this test and translated into values.

The initial slope of the stress-strain curve is used to determine the modulus of elasticity (E, unit: N/mm²). This value represents the stiffness of the material, whereas the maximum is considered as its tensile strength (σ_M unit: N/mm²). The stress strain curves of many thermoplastic polymers present an additional separate maximum, which can be lower than the tensile strength, as shown in Figure 2.3, but sometimes it is the maximum itself. This first maximum is called **yield point**, its coordinates are the **yield stress** σ_Y and the **yield strain** ε_Y. The tensile experiment ends with the failure of the probe at a point, that is characterized by the **strain at break** ε_B (elongation at break in %) and the **stress at break** σ_B (in N/mm²). In our example (Figure 2.3) the stress at break is identical with the **tensile stress**. Integration of the stress strain curve can yield the energy uptake for the whole deformation referred to the volume of the probe, representing the **toughness** of the plastic material. It is important to notice that mechanical values from the tensile test depend of the temperature as well as the elongation speed.

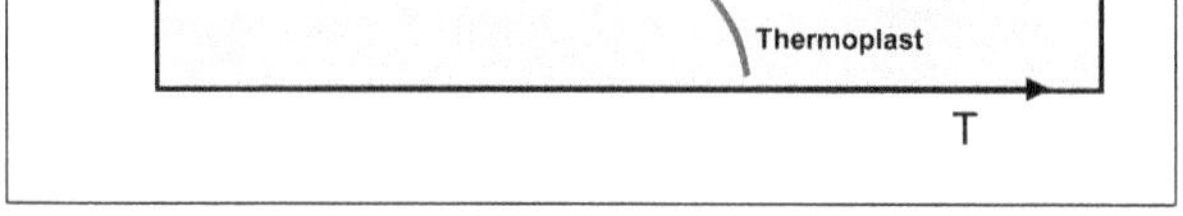

Figure 2.2: Dynamic Mechanical Analysis of plastics - simplified scheme (E': storage modulus, T: temperature)

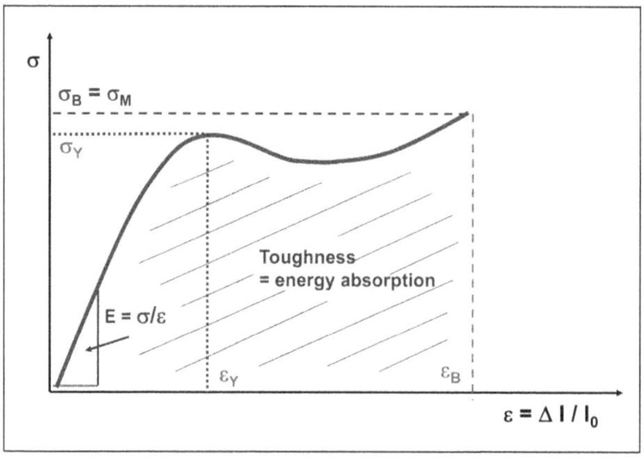

Figure 2.3: Tensile test - values

Table 2.3: Mechanical values of plastics, influence of various material variables *Source: database CAMPUS*

Plastic material	Tensile strength [MPa]	Modulus of elasticity [MPa]	Strain at break [%]	Notched impact strength a: Izod +23/-30°C [kJ/m²] b: Charpy +23/-30°C [kJ/m²]
PP	37	1450	>50	2/1.3 (a); 4.5/– (b)
PP-GF30	80	6200	3	26/24 (b)
PP+EPDM	18	1200 to 1500	>50	30 to 50/4 to 5 (b)
PS	45-55	3300	4	2/2 (a); 3 to 4/– (b)
ABS	45	2200 to 3000	30	37/22
PC	65	2200 to 2400	>50	–
PC+ABS	45	2100 to 2700	>50	–/16
PBT	55	2600	>50	–
PC+PBT		2200	>50	60/18
PA6 dry		2600	>50	
PA6 humide	85	1000	>50	
PA6-GF25 dry	160	8500		13/10 (b)
PA6-GF25 humide	100	5500		27/10 (b)
PPE+PA	60	2200	>50	65/30
PVC hard	75	3000	10-30	3.5/–
PVC-soft			>50	
UP	50 to 80	3000 to 4800		5 to 10
UP-GF28 lp (for SMC)	75	13,000		

It is a specialty of plastic parts that their mechanical properties can be controlled to a high extend by the material composition and processing parameters. Inorganic fillers help to increase moduli of elasticity and tensile strength values. Organic fillers increase toughness, carbon black pigments help to improve mechanical characteristics, electrical conductivity and resistance to light. Rigid plastics like hard PVC can be softened by the addition of plasticizers (esters, waxes). Tensile strength and stiffness can be increased many times over by the modification of polymers with fibres. The modulus of elasticity and the tensile strength are important values for lightweight construction, so as impact strength is for crash-performance. *Charpy* and *Izod* impact strength values of painted plastics, determined at 23 °C and -30 °C must not decline compared to the uncoated state. Table 2.3 shows some mechanical characteristics, as obtained by the tensile experiment according to ISO 527 and as documented in the data base CAMPUS Plastics. Additionally it shows the influence of elastification, reinforcement, softening and air humidity on the mechanical characteristics.

2.2.2 Thermal behaviour

Table 2.4 gives an overview of thermal characteristics of plastics, which reflects their softening and melting properties, the thermal conductivity and their thermal expansion.

Table 2.4: Thermal characteristics (glass transition and melting points, Vicat B50, thermal conductivity and coefficient of thermal expansion) [9-12]

Plasticmaterial	T_g/T_m [°C]	Vicat VST B 50 [°C] ISO 306	CTEα[K^{-1} 10^6] DIN 53752	Heat conductivity [W/(mK)] at 20 °C DIN 52612
PE-HD	< -100/125 to 135	60 to 65	120	0.35 to 0.51
PP	-20 to -10/160 to 165	92 to 94	150	0.22 to 24
PS	100	82 to 100	70	0.14 to 0.16
ABS	-85/95 to 105	90 to 100	60/110	0.17 to 0.18
PVC (hard)	70 to 80	70 to 90	80	0.16
PA6 (dry)	78	210 to 220	60	0.27
PC	145	160 to 170	60	0.21
PBT	45-60/220 to 230	180	60	0.20
PMMA	105 to 120	84 to 119	70	0.19
UP-20GF				0.12
UP-60GF				0.26
Iron			12	58
Aluminium			23	234
Quartz			1	10.5

Plastics obviously have a lower thermal resistance in comparison to metals. Glass transition (T_g) and melting temperatures (T_m) of plastics are significantly closer to ambient temperature compared to the melting points of iron (1538 °C) or aluminium (660 °C) (Table 2.4). The values for T_g and T_m can be obtained by determining either the E-modulus (DMA), the heat capacity (DSC) or specific volume as a function of the temperature. The exact values for T_g and T_m depend on the determination method and their specific parameters.

There are several material test methods in use to determine the softening temperature of plastics. In contrast to thermal analytical methods like DSC these material test methods do need larger probes of defined geometrical size. Measured values depend on the kind of the applied method, as the specimens being used differ in dimension and shape. One popular method of this range is the determination of the Vicat softening temperature.

The maximum application temperature for plastic parts depends on the duration of the heat load and on the kind of component part. The feasible drying and curing temperatures for coatings on plastics are limited by the thermal dimension stability of the plastic component parts. The thermal coefficient of thermal expansion (CTE) is an important characteristic for the estimation of the dimensional stability of plastic parts. Whereas in case of plastic substrates a low CTE value generally is desirable, for the substrate-coatings alloy this is mostly not the case. On the contrary it is important that the substrate and the coating do not differ to much in their thermal expansion, in order to avoid failures for example like delamination caused by shrinkage stress of the coating.

Due to their comparably low thermal conductivity, plastics are good thermal insulators. In the application process of coatings the thermal conductivity is a characteristic relevant for drying and curing. A coating material typically applied on steel, may get into a different curing state when applied and cured on plastics, as a result of the lower thermal conductivity and the resulting lower heat transport from the substrate to the coating. The orientation of metallic and pearleffect pigments in basecoats may also be influenced by the substrate-dependent thermal conductivity, by affecting the speed of solvent evaporation.

2.2.3 Solubility and swelling

The behaviour of polymers against organic solvents is described in the literature in different ways. One possible cause for this may be the ambivalent meaning of the term solubility for polymers. Depending on the view angle this characteristic can be regarded as to be something positive or negative.

As for industrial coatings the solubility of resins has been considered as quality characteristic. Before the age of synthetic polymers for coatings it was an important goal to achieve solubility of modified natural polymers resins, derived from sources like cellulose or latex. Solubility then was a decisive criterion for the suitability of polymers as binders.

However as for constructively used plastics, solubility has been regarded as something negative, affecting the resistance and work life of the objects which they are processed to. Under this aspect plastics are at a disadvantage compared to metals. Plastics have to fulfil demands of resistance against diverse media, solvents being one group of many others.

The scientific approach to describe solubility of polymers starts from the basic thermodynamic principle, that two components can form a solution if the free (Gibbs) energy of mixing is negative, according to *Gibbs* and *Helmholtz* (Equation 2.1):

Equation 2.1: $\Delta G_m = \Delta H_m - T\Delta S_m$

with ΔH_m = Enthalpy of mixing and ΔS_m = Entropy of mixing.

It can be expected that polymers do not get into solution under formation of heat that is why the value for ΔH_m in general will be positive. The value for ΔS_m also is positive, as the disorder of the whole system increases by mixing. The mixing will therefore take place in a volunteering way, if the entropy term $T\Delta S_m$ has a higher value than the enthalpy term ΔH_m.

Consequently, one option to promote the formation of a polymer solution is increasing the temperature of the mixture and therefore the term $T\Delta S_m$. Another possibility to realize the formation of a solution is to match solvent and polymer concerning their chemical structure, in a way that the mixture taking place only needs little heat of solution, ΔH_m. Dissolving a polymer in a solvent can be imagined as the penetration of small molecules inside and in between molecular coils, which means releasing intermolecular bonds between polymer chains and establishing new intermolecular bonds between solvent to polymer molecules. This can take place effectively if the intermolecular binding forces and cohesive strength of polymer and solvent are of similar dimension [13]. In practice this fact is transcribed with the catchphrase that similar substances dissolve each other.

A frequently used characteristic to describe this issue is the solubility parameter. Being equal to the square root of the cohesive energy per volume of a given substance, the solubility parameter of solvents can be determined by the heat of evaporation (unit: $(J/cm^3)^{1/2}$) see Equation 2.2.

Equation 2.2: $\delta = (E_c/V)^{1/2}$

with δ = solubility parameter, E_c = cohesive energy, V= volume

There are solubility parameters for solvents as well as for polymers. *Hildebrandt* in 1949 proposed, that the heat of solution ΔH_m of a polymer in a solvent should be described by the difference of solubility parameters of both components (Equation 2.3) [8].

δ*	Polymer	Solvent
32		
	Polyacrylnitrile	
30		
		Methanol
28		
	Polyamide 6.6	
26	Urea formaldehyd resin	
	Melamine formaldehyd resin	Ethanol
	Cellulose acetate	
24	Phenol formaldehyde	
	Polyvinyl acetate	2-Propanol, 1-Butanol, NMP
	Nitrocellulose	
22	Polyethylen terephthalate	
	Epoxy resin	2-Butanol
	Polyacrylic	
20	Polyurethane, polyoxymethylene	Cyclohexanone, acetone
	Polymethyl methacrylate	Butylglycol, butyl diglycol
	Polystyrene	MEK, diacetone alcohol
18	Polyisoprene (1,4-cis)	Xylene
	Polyvinylchloride	Solvesso 150
	Polybutadiene	
16	Polypropylene, polyethylene	
	Silicone	
14		n-Hexane
	Polytetrafluorethylene	

Figure 2.4: Scale of Hildebrandt-solubility parameters (in $(J/cm^3)^{1/2}$), classification of polymers and plastics*

* *often solubility parameter values are given in the old unit $(cal/cm^3)^{1/2}$ = 1 Hildebrandt; conversion: 1 $(cal/cm^3)^{1/2}$ = 2.046 $(J/cm^3)^{1/2}$* [14].

Equation 2.3: $\Delta H_m = V_m \, \Phi_1 \Phi_2 \, (\delta_1 - \delta_2)^2$

with ΔH_m : Mixing enthalpy (heat of solution)

V_m : Total volume of the mixture

Φ : Volume portion of the components in the mixture

δ : Solubility parameter

According to Equation 2.3 a good solubility of polymers in solvents can take place at best if their solubility parameters are quite similar, so that their difference is close to zero.

A comparison of solubility parameters of plastics and solvents is given by Figure 2.4.

One thing that can be read from Figure 2.4 is that for certain polymers like polyethylene, polypropylene, and silicones, apparently there is no adequate solvent.

While solubility parameters of solvents can be determined by their enthalpies of evaporation, in case of polymers this is not possible, because they cannot be evaporated without decomposition, due to their high molecular weight and high cohesive energy. However as for thermoplastic polymers, in principle solubility parameters can be obtained by determining the intrinsic viscosity of a polymer solution, using different solvents with known solubility parameters (δ-values). The solubility parameter of the polymer then can be considered as to be identical with the one of that solvent causing the highest intrinsic viscosity [14]. The solubility parameter of cross-linked polymers can be determined by investigating their swelling, as shown in Figure 2.5.

At first sight the concept of ***Hildebrandt*** seems to be a reasonable model to describe the miscibility of polymers in solvents. But it has some limitations. For example the solubility of polymers declines with increasing degree of crystallinity and there are only ***Hildebrandt*** parameters for amorphous polymers. At next the solubility parameter of a given polymer increases with its molecular weight, so it is no characteristic with an exact and constant value. Comparing and subtracting ***Hildebrandt*** parameters of polymers and solvents may provide a first orientation for questions of solubility. While for non polar substances it often fits to the experience, this concept generally fails for substances cohering dominantly by

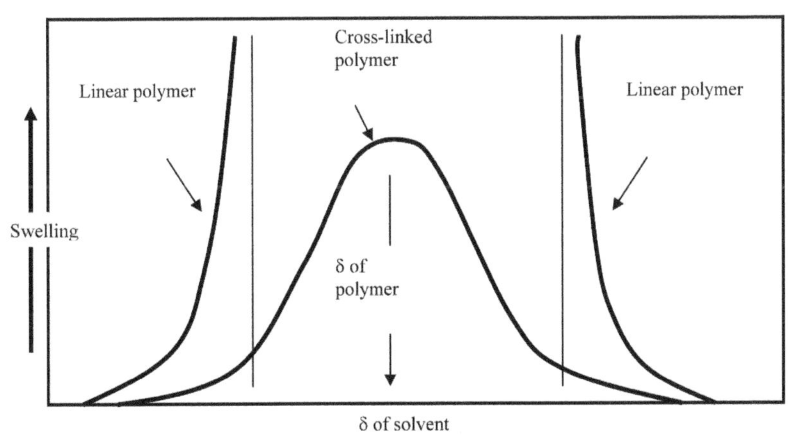

Figure 2.5: Swelling of linear and cross-linked polymers as a function of the solubility parameter, according to [7]

dipole-dipole-forces or hydrogen bonds. It could be shown that cohesive forces are at best to be described as a sum of intermolecular attractions differing significantly in magnitude. According to *Hansen* the solubility parameter is composed of three parts: one each for hydrogen bonds, dipole-dipole-forces and dispersive forces (Equation 2.4).

Equation 2.4 (*Hansen* solubility parameter): $\delta = (\Delta_D{}^2 + \delta_P{}^2 + \delta_H{}^2)^{1/2}$

δ_D for dispersive forces

δ_P for dipole dipole forces and

δ_H for hydrogen bonds.

The concept of *Hansen* has found conformation in practice. A given polymer should dissolve in a solvent if their *Hansen* parameters have similar values (Tables 2.5 and 2.6). Investigations have shown that the similarity of δ_P is of particular importance.

The solubility of a polymer in different solvents can be estimated by regarding the difference of their *Hansen* parameters, in analogy to the *Hildebrandt* concept. A solubility then is considered as good, if this difference is minimal or does not exceed a random value R (Equation 2.5, P: polymer, L: liquid/solvent) [18].

Equation 2.5: $R = \{ [4 (\Delta(L)_D - (\Delta(P)_D]^2 + [(\Delta(L)_P - (\Delta(P)_P]^2 + [(\Delta(L)_H - (\Delta(P)_H]^2 \}^{1/2}$

Those values R are listed as well as *Hansen* parameters for a range of polymers, including a lot of binders for coatings, which in many cases have to have a good solubility in paint solvents. Unfortunately there is a lack of *Hansen* parameters for many of those polymers serving as substrates for coatings. This is a limiting factor for the *Hansen* concept, which on the other hand can provide valuable insights into the solubility of polymer classes, moreover as there are *Hansen* parameters for most of the relevant coating solvents.

Besides the scientific approaches to describe solubility of polymers in solvents, there are also some evaluation schemes based on practical experience. These presentations rate plastic materials according to their ability to withstand the attack of solvents. So they classify plastic materials according to their inertness to solvents. Especially in case of plastic materi-

Table 2.5: Hansen parameter of solvents in the unit $[(J/cm^3)^{1/2}]$ [15]

	δ $[(J/cm^3)^{1/2}]$	δ_D $[(J/cm^3)^{1/2}]$	δ_P $[(J/cm^3)^{1/2}]$	δ_H $[(J/cm^3)^{1/2}]$
n-Hexane	14.9	14.9	0	0
o-Xylene	18.1	17.8	1.0	3.1
Cyclohexanone	19.6	17.8	6.3	5.1
Methyl ethyl ketone	19.0	16.0	9.0	5.1
Acetone	20.0	15.5	10.4	7.0
Butyl acetate	17.4	15.8	3.7	6.3
2-Propanol	23.5	15.8	6.1	16.4
1-Butanol	23.1	16.0	5.7	15.8
Isobutanol	22.7	15.1	5.7	16.0
Butyl glycol	20.8	16.0	5.1	12.3
Butyl diglycol	20.4	16.0	7.0	10.6
n-Methyl pyrrolidone	22.9	18.0	12.3	7.2
Water	47.8	15.6	16.0	42.3

als that do not have to be coated or glued, the resistance against solvent attack seems to be a useful quality criterion. Looking at plastic materials that have to be coated, it should be respected that an interaction of the substrate with solvents may positively influence coating adhesion. On the other side this interaction must not so go so far as it results in softening, dissolution or formation of stress cracks.

If a plastic material is under residual stress, then specific solvents may relieve this stress spontaneously by slightly softening and dissolving the polymer surface, leading to stress cracks.

The presently available information about practically orientated resistances of plastic materials to coating solvents do not give a sharp picture. Publications dominantly contain results of solvent groups [9, 19-21]. Moreover these results are mainly not expressed in data but in words, reflecting an assignment to an individual evaluation scale (for example: resistant, conditionally resistant, none resistant). No wonder that those results of different authors cannot be matched to each other. Naturally they also underlie fluctuations resulting from the large variety of material types and non-identical test conditions.

The overview in Table 2.7 is more or less a compromise derived from the various existing information about solvent resistance of plastics. Being aware that a compromise of such a broad variety of facts is only possible to a very limited extent, this table of course does not claim to be a scientific characterization but is more an attempt to respect different facts of the above cited sources in a somewhat balanced out way.

The following trends can be concluded:

- Styrene containing plastics in general are less resistant to solvents, with the exception of some aliphatics and alcohols.

Table 2.6: Hansen parameter of plastics and polymer binders in the unit $[(J/cm^3)^{1/2}]$ [7, 15-17]

	δ $[(J/cm^3)^{1/2}]$	δD $[(J/cm^3)^{1/2}]$	δP $[(J/cm^3)^{1/2}]$	δH $[(J/cm^3)^{1/2}]$	Ref.
Polyethylene (PE)	17.3	17.3	0	0	[16]
Polypropylene with EPDM (TPO)	18.4	18.0	3.2	1.9	[17]
Polytetrafluorethylene (PTFE)	12.7	12.7	0	0	[16]
Polybutadiene (BR)	18.8	18.0	5.1	2.5	[7]
Polyisoprene (NR)	18.0	17.4	3.1	3.1	[7]
Polystyrene (PS)	20.1	17.6	6.1	4.1	[7]
Styrene butadiene copolymer (SBR)	18.9	17.4	2.9	6.8	[15]
Polyvinyl chloride (PVC)	22.5	19.2	9.2	7.2	[7]
Polyvinyl acetate (PVA)	23.1	19.0	10.2	8.2	[7]
Polymethyl methacrylate (PMMA)	23.1	18.8	10.2	8.6	[7]
Polyethylen terephthalate (PET)	–	–	3.5	8.6	[16]
Polyacrylnitrile (PAN)	25.6	18.6	16.2	6.8	[16]
Polyether urethane (PEU)	20.8	17.4	5.1	10.2	[16]

Table 2.7: Solvent resistance of some plastic materials at 23 °C

Polymer	Aliphatics	Aromatics	Ketones	Esters	Glycol ethers	Alcohols
PS	0	–	–	–	+	+
SB	–	–	–	–	0	+
ABS	+/0	0/–	–	–	0	+/0
PC	+	0	0/–	0/–	+/0	+
PMMA	+	0	0/–	0/–	–	0
PBT	+	+/0	0	0	+	+
PA	+	+	+	+	+	+
PE	+/0	+/0	+	+	+	+
PP	+/0	+/0	+	+	+	+
PVC-U*	+	+/0	0/–	0/–	+	+
PVC-P*	0	–	–	–	0	0
POM	+	+	+	+	+	+
PPO	+/0	0	–	0	+	+
UP**	+	0/–	0/–	0/–	0	+/0
PUR**	+	+	+/0	+	+/0	+/0

+ : resistant; 0 : conditionally resistant; - : not resistant
**PVC-U: Hard-PVC, PVC-P: Soft-PVC*
***UP as duroplast and PUR as elastomer*

- Polyolefins, especially polyethylene, are mainly resistant, with slight curtailments of hydrocarbons.
- POM and PA are resistant against all solvents.

In the first instance, this information is only of orienting value and it is necessary to check on the specific case in practice. It does not make sense to rely too much on information referring only to classes of solvents and plastic materials. Solubility depends also on material parameters like: molecular mass, branching degree, network density, content of fillers and fibres, as the way of processing, that the material might have gone through. Plastic foams for example have a low wall thickness and therefore are less solvent resistant than polymers that have not foamed. At the end it is advisable to perform experiments.

A particular kind of swelling is water absorption. Polyamides are known for their tendency to take up and give off water, depending on the air humidity. The thermal and mechanical properties of polyamides therefore are dependent of their water content, causing a softening effect: Glass transition temperature, stiffness and tensile strength decline with increasing water content. Even more severe concerning coating performance can be the formation of pops and pinholes caused by incomplete evaporation of water during the drying process, whereas it is appropriate to define a humidity content of polyamides and its blends to be coated. This value should reflect the properties of coating materials as well as the application and drying conditions. The substrates then have to be conditioned long enough in order to secure an appropriate condition. It seems not to be necessary to get polyamides completely dry, as a water content of 1.6 % in case of blends of PA and PPO has been proven to be feasible for a process-safe production of coated automotive build on parts.

2.2.4 *Electrical properties*

The electrical behaviour of plastic materials can be described by different characteristics. One of those important characteristics is the specific on-resistance, expressed by the symbol ρ_D. This value can be obtained by the electrical resistance Ω of a plastic probe with the thickness d and the surface area A, being placed between two plate electrodes (Equation 2.6).

Equation 2.6: $\rho_D = \Omega\ A/d\ \ [\Omega\ cm]$

Metals with excellent electrical conductivity like aluminium, copper or silver have ρ_D-values of 2.8 10^{-6} to 1.6 10^{-6} Ω cm, whereas plastics are typical non conductors with values between 10^8 and 10^{18} Ω cm (Table 2.8). Plastic substrates for coatings do not conduct electrical current, so they are typical non-conductors, especially that counts for non-polar polymers like PTFE, PE or PS. By adding carbon black, the electrical conductivity can be increased significantly. Plastic materials containing electrically conductive fillers show ρ_D-values from 10^7 to10^{-2} Ω cm, for example the inline-coat-able PPO+ PA-blend: 10^5 Ω cm.

One Siemens is the reciprocal unit of one Ohm. $1/\Omega = 1$ S. Therefore the correct unit for specific conductivity is S/cm. The specific electrical conductivity of intrinsic conductive polymers is specified using this unit, values being in the range of 10^{-5} to 10^3 S/cm. Copper has a specific conductivity of 6.5 10^5 S/cm, aluminium of 3.7 10^5 S/cm. Intrinsic electrically conductive polymers are for example polyaniline, polyacetylene und polythiophene [22].

Table 2.8: Specific on-resistance of some materials [11]

Material	Specific on-resistance ρ_D [Ohm cm]
Fluor polymers	10^{18}
PVC	10^{15} to 10^{16}
PS	10^{18}
Polyolefins	10^{18}
PMMA	10^{16} to 10^{17}
PC	10^{18}
PA (dry)	10^{16} to 10^{17}
POM	10^{16}
EP	10^{16} to 10^{17}
UP	10^{16} to 10^{17}
PA/PPO el. cond.	$< 10^5$
ABS + 15 % CF	10^4
Glass	10^{13} to 10^{14}
Aluminium	$2.8 \cdot 10^{-6}$
Copper	$1.7 \cdot 10^{-6}$
Silver	$1.6 \cdot 10^{-6}$

The extremely low electrical conductivity of standard plastics is significant for the functionality of plastic as a coating substrate, basically in two aspects:

- Plastic surfaces can charge electrically and the charges cannot be removed without external help. Possible consequences are the attraction of dust and optical impairment of the coated surface by dust particles.
- Without a sufficient conductivity of the substrate surface it is difficult to make use of electrostatically supported paint application and its benefit to increase transfer efficiency. In practice these difficulties can be overcome by the application of an electrical conductive primer, in order to bring the part to state where it is suitable for ESTA-application. Another option is to add carbon black pigment to the whole plastic bulk. This pigment can also function as a UV-absorber and protect the polymer from sunlight that attacks and degrades most plastics over time. Moreover polymer modification with carbon black pigment is a way of reinforcement and used since a long time in the tire industry. Whether plastic materials with significant electrical conductivity are suitable as substrates in industrial coating processes depends on process stability, electrical conductivity and the price. Up to now they have not found extended market penetration in coating technology.

2.2.5 Surface characteristics

Two characteristics having a strong influence on the optical and functional performance of coatings are

- surface structure and
- surface tension.

2.2.5.1 Surface structure

The topography of plastic surfaces can influence optical properties and adhesion of coatings.

As known from experience the waviness and roughness of a substrate surface not always can be levelled by a coating, but on the contrary can be telegraphed to the coating surface and thereby even be increased. Figure 2.6 shows photographs of coating surfaces on different substrates.

With a given plastic substrate the surface structure of a moulded part is strongly dependent on the quality of the used processing tool and the processing conditions. Also the substrate quality can get changed by a surface pre-treatment. For example sanding of plastic surfaces usually results in roughening, having a positive influence on coating adhesion. When the processing generates moulded parts with a rather rough surface, it usually can be levelled by sanding. On the other hand sanding can involve risks like condensation blisters, being formed in sanding marks under humid conditions. Sanding dust and fibres may adhere to the surface and can cause coating surface defects.

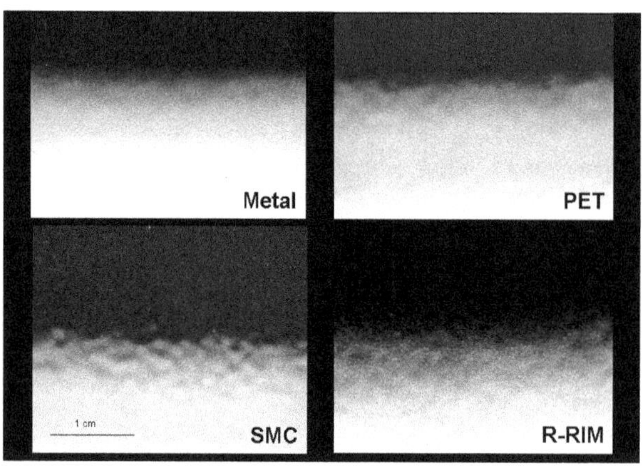

Figure 2.6: Photographs of coating surfaces on different substrates (photo from [23])

Roughness measurements of plastic surfaces are mostly performed with a surface profilometer in analogy to measurements on metal. The feasibility of this method is limited by the low hardness of the substrate, because it is one prerequisite of mechanical profilometry that the stylus instrument being used must not penetrate into the substrate.

2.2.5.2 Surface tension

The wetability of a plastic material depends on its surface tension. This surface tension is a surface related energy with the unit Joule/m^2 or N/m, and its meaning is the tendency of the surface to reduce its area, or in other words how difficult it is to enlarge the surface area. The fundamentals of wetting theory are described profoundly in literature, so that a brief summary should be sufficient here [24, 25].

As a consequence of the definition above, solid materials with a comparable high surface tension are more easily to wet than those with low surface tension. The reason for this is, that by getting wetted, the free enthalpy of the whole system is minimized by decreasing the solids surface area, being identical with the interfacial area of solid/air. The wetting of a surface by a liquid can be described by the contact angle model according to *Young*, as shown in Figure 2.7. The liquid droplet spreads on the surface with a contact angle depending on the interfacial tensions between the associated solid, liquid and gas.

Good wetting is characterized by a small contact angle of the liquid droplet. The value of this contact angle not only depends on the surface tension of the substrate, but also of the liquid. The *Young* equation represents the relationship between contact angle and interfacial tensions (energies), see Equation 2.7.

Equation 2.7: $\gamma_{sg} = \gamma_{sl} + \gamma_{lg} \cos \theta°$

If the contact angle Θ is 0° then cos Θ has its maximum value of one, whereas at 90° it adopts the value of zero. According to *Young*'s equation, low contact angles Θ and therefore high values for cos Θ are resulting if the surface tension γ_{sg} of the solid to be wetted is rather high and the surface tension γ_{lg} of the wetting liquid is rather low. However the wetting model of *Young* is only valid under the following boundary conditions:

- the substrate surface has to be smooth and homogeneous
- the substrate must not be swollen by the wetting liquid

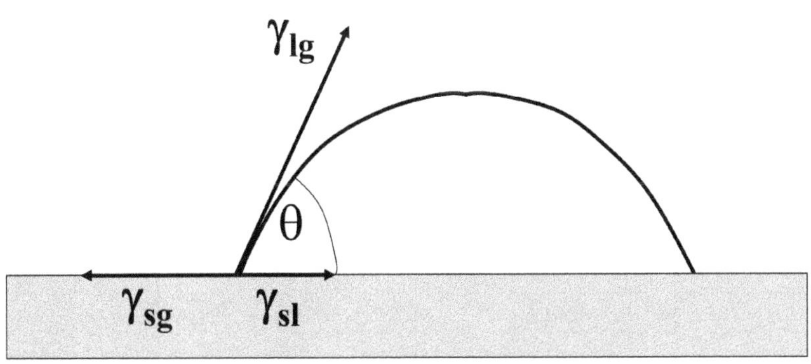

γ_{sg}: *interfacial tension solid/gas* γ_{ls}: *interfacial tension solid/liquid*

γ_{lg}: *interfacial tension liquid/gas* Θ : *contact angle*

Figure 2.7: Wetting of a solid - contact angle model according to Young

As can be seen in *Young*'s equation, those liquids with low surface tension do generally wet surfaces more easily. The maximum surface tension of a liquid, barely allowing to wet a given solid surface, is called **critical surface tension**. Of course the value of this characteristic depends not only on the kind of substrate but also on its surface condition, being possibly the result of additional factors like surface-near variations in blend composition and surface chemistry properties. The critical surface tension of a substrate can be obtained according to *Zismann* by plotting cos Θ-values of liquids on the surface of this substrate across their known values of surface tension γ_{lg} and extrapolation of γ_{lg} against $\cos \Theta = 1$ [26]. For many practical issues it is sufficient just to get an estimation of the wetability of a plastic surface. For those cases a less scientific version of the *Zismann* method might be probable. By this method it is possible to estimate the surface tension of the substrate by correlation with the surface tension values of test inks applied with a brush. The surface tension value of the one ink spreading over the surface for more than two seconds may be regarded roughly as the substrate's surface tension. In fact this value should not be considered as an absolute characteristic but more as a relative value being close to the critical surface tension. Test inks being available on the market are mixtures of ethanol and water.

With respect to the different intermolecular forces, determining the cohesion energy of substances, the surface tension can be splitted into a polar part γ^p and a disperse part γ^d (Equation 2.8) [27]. The polar forces represent a combination of dipole-, induced dipole-, and hydrogen bonding forces.

Equation 2.8: $\gamma = \gamma^d + \gamma^p$

Disperse and polar parts of the surface tension are well documented for many liquids and polymers [7]. In case of solids they are obtained by measuring the contact angle with five liquids of known polar and disperse parts of surface tension [24]. Table 2.9 shows a survey of surface tension values of plastics.

Similar concepts using acid-base models have been developed by *Fowkes* and *van Oss* [28-31].

Wetability and coating adhesion

Today it is accepted that good wetting of a substrate is one requisite for coating adhesion. In the 19th century *Dupré* developed a thermodynamic theory of adhesion. He correlated the work of adhesion with the interfacial energies between the participant phases (Equation 2.9). By inserting *Young's* equation into *Dupré's* equation the direct connection

Table 2.9: Surface tension of some plastic materials in comparison to-hexane, ethanol and water [7]

Material	Surface tension, overall [mN/m]	Surface tension, disperse part [mN/m]	Surface tension, polar part [mN/m]
PE linear	35.7	35.7	0
PE branched	35.3	35.3	0
PP isotactic	30.1	30.1	0
PET	44.6	35.6	9.0
PMMA	41.1	29.6	11.5
PTFE	20.0	18.4	1.6
n-Hexane	18.4	18.4	0
Ethylene glycol	48.0	33.8	14.2
Water	72.8	21.8 ± 0.7	51.0

between wetting, expressed by the cos value of the contact angle, and adhesion becomes obvious (Equation 2.10).

Equation 2.9: $W_a = \gamma_{lg} + \gamma_{sg} - \gamma_{ls}$

Equation 2.10: $W_a = \gamma_{lg} (1 + \cos \theta)$

A problem arising from this result is that one might presume that adhesion can be maximized by wetting alone. The common observation that coating adhesion on certain plastic substrates is improved by surface activation apparently has led to the popular assumption that this might be solely reduced to an improved wetability. Indeed the wetability of polymers by coatings often is improved significantly by surface activation methods.

It has been proven several times that plasma activation or flame treatment of polyolefin's results in an increase of the polar part of surface tension [32, 33]. Nevertheless a continuous correlation of coating adhesion with surface tension of plastics could not be proven [33, 34]. Moreover it has not succeeded yet to calculate adhesive strength values from interfacial energies. It might appear evident that a high polarity of a plastic surface promotes wetability and coating adhesion. But drawing conclusions from this to a straight cause-effect correlation does seem to result in a too simple picture. Perhaps this is because of the fact, that *Dupré's* theory treats adhesion as a pure matter of thermodynamics, based solely on wetting. Thereby prerequisites are attributed to adhesion which is not given in reality, for example, complete reversibility of wetting. It seems to be unlikely that coating adhesion can be controlled solely by the wetability of the substrate. For sure wetting is a necessary prerequisite for establishing adhesive bonds, but not the only mechanism being causing adhesion.

2.3 Successful coating of plastics

2.3.1 General remarks

There are five fields of influence deciding on the success of coating plastics:

- Substrate material
- Plastic processing
- Surface pre-treatment
- Coating material
- Coating application

Factors that have influence on the coatability of plastic materials are: kind and status of the polymer, fillers, fibres and additives. As with coating formulations, the percentage of all ingredients is an important factor for the properties of plastic materials, not least because of possible consequences for surface properties and interaction with the attached coating layer.

The geometry of the part to be coated, processing parameters, storage and aging as well as surface pre-treatment also have a powerful influence on the result of the coating process.

In addition to substrate related factors, an equal important influence on the coating process results from coating material and application design. Coated plastics can be regarded as polymer laminates, being composed of a polymer substrate covered with a multiple layer polymer composite. Basically it is possible that interactions between the coating body and the substrate lead to an exchange of substances. However the migration of ingredients is hard to track and prove.

The success in coating objects of plastic depends also strongly on various parameters of the painting process. Curing parameters like temperature and time do have an influence on dimension stability of the part as well as film integrity. Positioning of the parts to the paint application robot can also affect film properties, surface quality and colour matching. Many principles in paint application, paint shop lay out, curing concepts, paint supply and disposal can be taken from general experiences in other fields of industrial coatings and be transferred to the coating of plastics. To coat a moulded part of defined shape, size, substrate and surface condition with low reject rate at moderate costs mostly affords an individual planning of the whole coating process. In most cases the coating process for a specific object is not designed with respect to a new, tailor suited paint shop, that would allow an optimum of parameters being adapted to the object and intended to use paint material. Mostly the coating process has to be carried out in an existing paint shop, including established paint application devices, and the resulting constraints in combination with specifications can strongly limit degrees of freedom concerning coating material and process parameters. There are many possible combinations of coating material and application parameters as well as objects to be coated, that make the coating of plastic parts an exciting business.

2.3.2 Substrate dependent influences

In which way does the substrate material influence the coating layer's quality? At first the kind of used base resin plays a decisive role, because it mainly determines the material properties. As shown in Table 2.1, among the plastic substrates to be coated, virtually all classes of polymer groups are represented. Regarding a polymer type, producers do often have more than one version of that type in their portfolio, differing in viscosity and other technical characteristics. Additionally the variety of polymer variants of each type available on the market is increased by the number of producers. Plastic materials are not only composed of polymer resins, but in most cases contain additives and other aggregates, and some of them are well known from coating formulation:

- Fillers
- Fibres
- Softeners
- Colourants
- Antioxidants
- Stabilizers
- UV absorbers
- Lubricants
- Internal release agents
- Flame resistance additives

The plastic material's composition has to respect the various demands of moulded part's and semi-finished product's processing and specific purposes. The additives do have a tremendous influence on the material properties. As valuable as they are for optimizing the technical profile of plastics, additives can also affect severely coating properties.

Internal release agents

Realizing that the main issue of coatings on plastics is adhesion, all of the substrate's ingredients do have to be regarded attentively. This is especially true for release agents. If they are not carefully removed by a cleaning pre-treatment, then adhesion failure is a frequent consequence. Release agents can be detected by methods like TOF-SIMS, ATR or FT-IR.

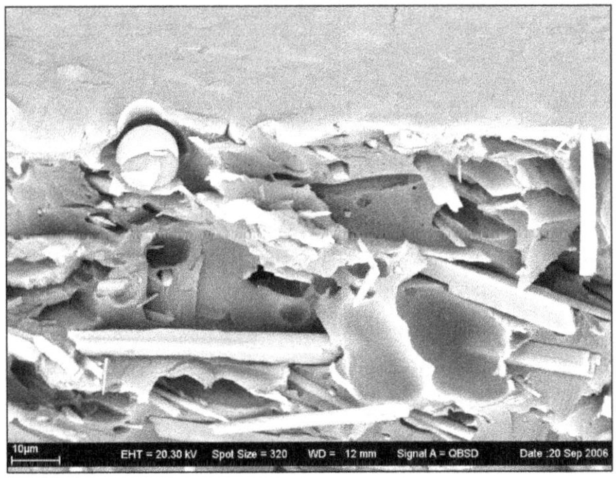

Figure 2.8: Scanning electrode microscope image of a component from PUR-RIM in the cross section

Reinforcements

The inner life of reinforced plastics like RRIM or SMC as represented by the electron microscope appears as rather porous and heterogeneous mixture. Besides a high content of, filler particles and beads as well as the polymer matrix there also are interstices and hollows (Figures 2.8 and 2.9).

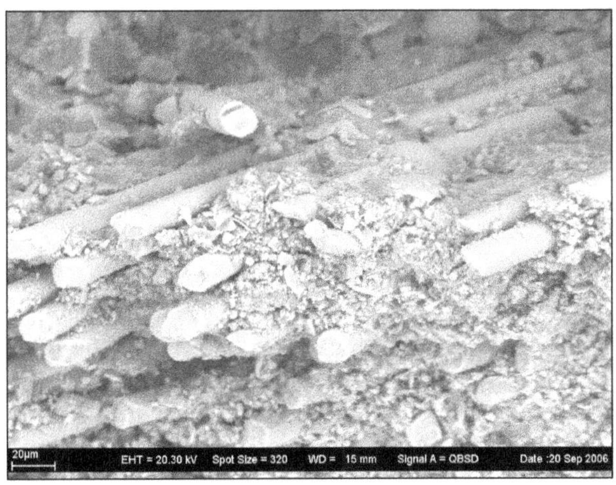

Figure 2.9: Scanning electrode microscope image of a component from SMC in the cross section

As a consequence from insufficient wetting of fibres and the presence of volatile ingredients, several coating failures are observed. SMC tends to develop shrink holes and out gassing. As for fibre reinforced substrates dispensing with the primer can result in fibre marks being visible in the coating. Door handles and grip rails from fibre reinforced polyamides tend to telegraph these markings up to the clear coat. Flame pre-treatment can even aggravate this effect.

Softeners and other organic additives

Some plastics like soft-PVC contain softeners, which can migrate to the surface. If such a plastic is going to be coated, softeners can accumulate in the interface between coating and substrate, weaken adhesion, so that delamination can take place. An ongoing migration of softeners into the coating layer can result in declining resistance properties of the coating surface.

Additives like anti-oxidants, stabilizers, lubricants and dispersing agents can also migrate to the surface and weaken adhesion and cohesion of the coating layer [35].

All these compounds are quite easily detectable on plastic surfaces by analytical methods like infrared spectroscopy using reference spectra. The tendency to migration can be examined by tempering experiments. The plastic part has to be cleaned first, then heated up and washed with a hydrocarbon solvent. After distilling the solvent off, the chemical composition of the remaining residue may be characterized using spectroscopical methods.

Water and other volatile ingredients

Polyamides and polyamide blends like, for example PPO+PA take up water during storage and emit it when getting heated up. This fact plays a role in coating processes, because the water remaining in the coating after flash off can result easily in popping and pinholes. Careful conditioning the plastic component in a defined climate prior to painting as well as curing under reduced air humidity can provide control of this problem. In polyamides water acts as a softener, causing decrease of stiffness and tensile strength.

Emission of remaining monomers or other volatile organic ingredients can react, as in the case of SMC, in popping and pinholes, especially on the edges. Apart from technical aspects emissions from plastics are a safety issue. Plastic emissions can endanger the health of persons exposed to them. Therefore an important criterion for the evaluation of plastic materials for automotive interior purposes is the tendency to emit chemical compounds.

Table 2.10: Coatabilityof plastics

Plastic	Abbreviation	Thermal stability [°C]	Difficulties
Polypropylene	PP, TPO	80 to 110	No adhesion without activation (depending on rubber content*)
Polyethylene	HD-PE LD-PE	60 to 80	No adhesion without activation (depending on rubber content*)
Polystyrene, expanded Polystyrene	PS, EPS	50 to 70	Solvent attack, stress cracks (esters, ketones, aromatics)
Acryl-/butadien-/styrol-copolymer	ABS	70 to 80	Solvent attack, stress cracks (esters, ketones aromatics)
Polycarbonate	PC	90 to 110	Solvent attack, stress cracks (esters, ketones aromatics), alkali-sensitivity
Polybutylene terephthalate	PBT	130 to 160	Adhesion not ensured
Polyamide	PA	90 to 150	Popping (pinholes) of water
Polymethyl methacrylate	PMMA	65 to 80	Solvent attack, stress cracks (esters, ketones aromatics)
Polyvinyl chloride	PVC	60	Soft-PVC: release of softeners Hard-PVC: adhesion failure, solvent attack, stress cracks
Phenol-, urea-, melamine resins	PF, MF, UF	110 to 160 (depending on filler content)	none
Polyurethane expanded and reinforced	PUR, RIM, RRIM	80 to 180	Porosity release agents
Unsaturated polyesters, reinforced	UP, SMC, BMC	180	Porosity, release agents, slip additives
Epoxy resins, reinforced	EP	160 to 220	Release agents

** Coating adhesion in generlly is better with higher rubber content*

The material parameter presented thus far may allow for some conclusions concerning the suitability of plastic types for use as coating substrate. As those conclusions are drawn from specific experience, they do only have a limited universal validity. Under the condition that exceptions are always possible, the Table 2.10 on page 40 may help to provide some orientation concerning the coatability of plastic materials.

2.3.3 Influence of processing, storage and transport

2.3.3.1 Cause-effect relationships

The before mentioned influence of the substrate structure on coatings surface quality can easily be observed with SMC, RRIM and some reinforced thermoplastic materials. Within each group of polymer substrates the material derived influence on coating layers properties is additionally superimposed by the influence of processing. These cause-effect relationships are represented by the following issues, which are further outlined (Figure 2.10).

Surface condition

Tool quality has a big influence on surface roughness and structure. Automotive applications afford class A-surfaces, which may allow an optional dispense with the primer and transfer of primer functions to the basecoat.

The melt flowing into the moulding tool can leave permanent flow marks. Homogeneity and the rheological profile of the melt during the cooling stage are determinants. A typical pattern of flow marks is observed as a tiger skin appearance. Often moulded parts show weld lines and gate marks. These zones mostly are weak points of mechanical strength and solvent resistance. Coatings may, even if appropriate in any other aspect, optically intensify gate marks and weld lines. In such cases the solvent composition of the coating should be adjusted with respect to the substrate's solvent sensitivity without sacrificing adhesion. As known from experience, flow lines can be created in different intensity by variation of injection moulding parameters. With a given plastic material meeting the specifications it should be possible to work out an appropriate set of processing parameters, which allow the production of parts free from flow lines and marks. On the other hand a lot of other technical demands have to be fulfilled and productivity calls for short cycle times, resulting in constraints that may make it difficult to produce perfect surface quality.

Processing, handling, storage and transport involve risks of surface contamination with external release agents, grease, sweat, oil and dirt. Furthermore processed parts can be mechanically damaged.

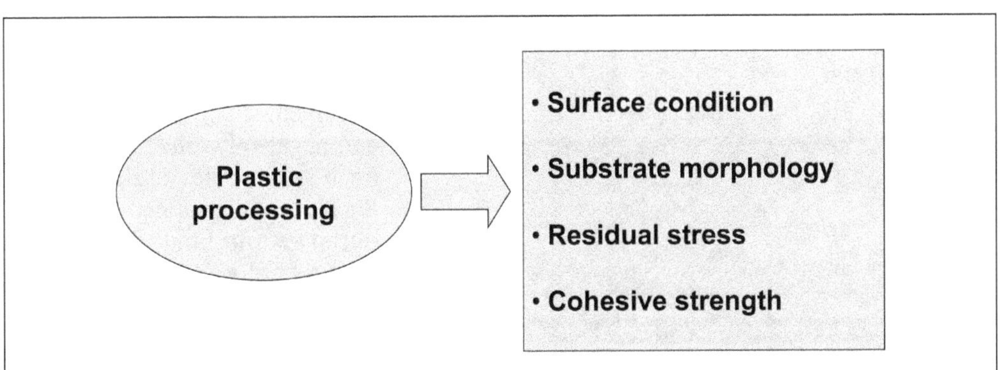

Figure 2.10: Plastic processing and effect on component's characteristics

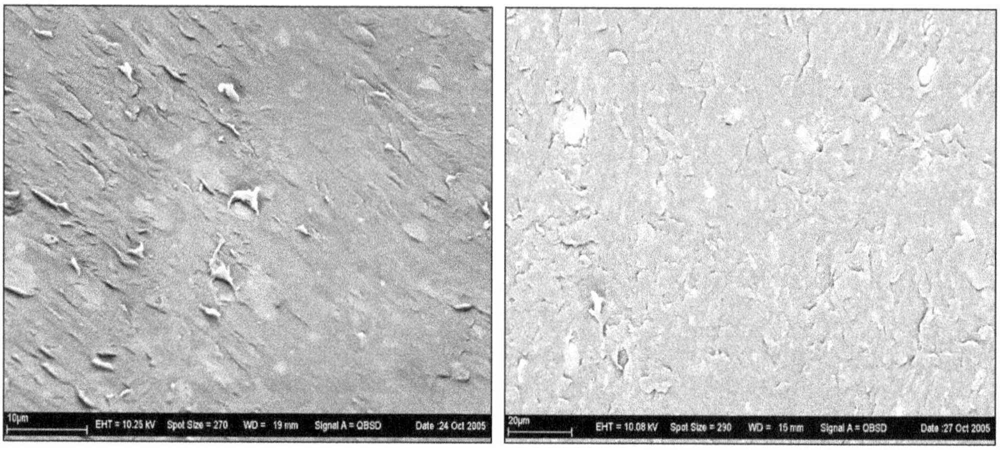

Figure 2.11: SEM photograph of bumper fascia front and side from TPO

The moulded component's geometry also has a tremendous influence on coating quality, because it affects both processing and coating cure. Mould flow in the tool can result in local specific differences of material bulk characteristics as well as coating performance. An example of local differences in surface structure of injection moulded plastic parts is presented in Figure 2.11, showing electron micrographs of bumper fascia front and side regions.

Naturally it cannot be excluded that local specific surface and material properties may cause local dependent differences in the interaction of the substrate with the coating.

The efficiency of film forming and cure both are influenced by the part's three dimensional shape. Coating application and curing conditions in a painting line have to be carefully matched to the individual design of the component, in order to achieve maximum of process stability by coating film properties as even as possible. UV curing of coatings on three dimensional shaped components is a real challenge, as curing can only take place where radiation gets to, which is hardly the case in shadow zones. It is fairly safe to state that local conditions of pre-treatment, coating application and curing should generally fit to the shape of the objects to be coated. If for example adherent water from power wash treatment is not properly removed before the coating application is to take place, then formation of blisters under humidity impact can be the result.

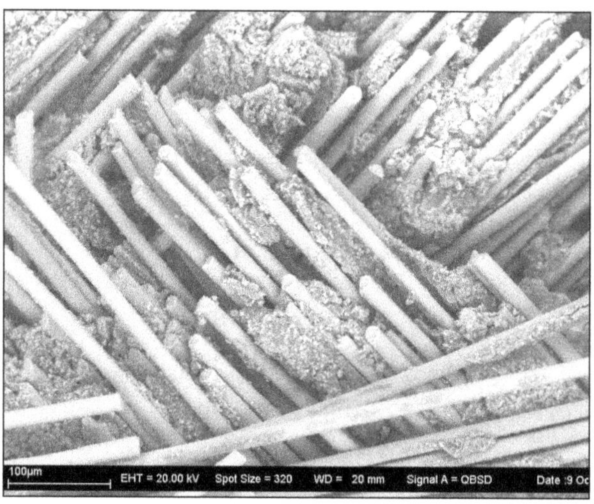

Figure 2.12: SEM photograph of a cross section of a SMC component

As mentioned before, the reinforcement of polymers with fibres can negatively affect the appearance of coated plastic parts. Surface quality is not only influenced by the kind, content and orientation of fibres, but also by the polymer type and processing viscosity. Wetting of the fibres by the polymer affect not only cohesive strength of the composite, but also surface homogeneity and the occurrence of surface defects. Materials like SMC with high loadings of long fibres often have a defined orientation of fibres (Figure 2.12).

Processing parameters like temperature and injection pressure strongly drive the melt viscosity. Therefore they have an influence on wetting of the fibres by the polymer. The mobility of short fibres in a polymer melt declines with increasing polymer viscosity, as for example with glass fibre reinforced polyamides. Consequently fibre orientation strongly depends on processing parameters.

Substrate morphology

The shear stress of a polymer melt can affect coating adhesion in different ways. In some cases excessive shearing involves the formation of a flaky surface near substrate region, causing coating delamination. But also positive aspects of shear stress by injection moulding have been reported. In the case of TPO, the connection between processing parameters, substrate morphology and effect on coating adhesion is well documented [36-38]. Short cycle times afford high injection pressures, and the resulting high shear forces cause the formation of a specific layer morphology of the produced parts (Figure 2.13).

Thickness and compatibility of these layers not only impact the substrate's cohesive strength, but also play an important role in the homogenization of polypropylene with rubber as well as of crystallinity. Process parameters are: melt viscosity, ratio of polyolefin to elastomer, cooling rate and cooling temperature. Coating adhesion on TPO is mainly directed by the consistency of the substrate phase close to the interphase, as interdiffusion of polymer chains depends on the plastic's composition and crystallinity. It is generally believed that polymer chain diffusion into the polypropylene matrix is promoted by surface near rubber. This idea fits to the fact that on compression moulded TPO there has not been found any coating adhesion, in contrast to injection moulded TPO.

Internal stress

There are several potential origins causing internal stress in a plastic part. Injection moulding with too short cycle times can cause so called cooling stresses. Those are partly a result from low thermal conductivity of the plastic material. Fast cooling lets freeze the polymer melt close to the tool surface with a faster rate than in deeper regions, where the polymer has

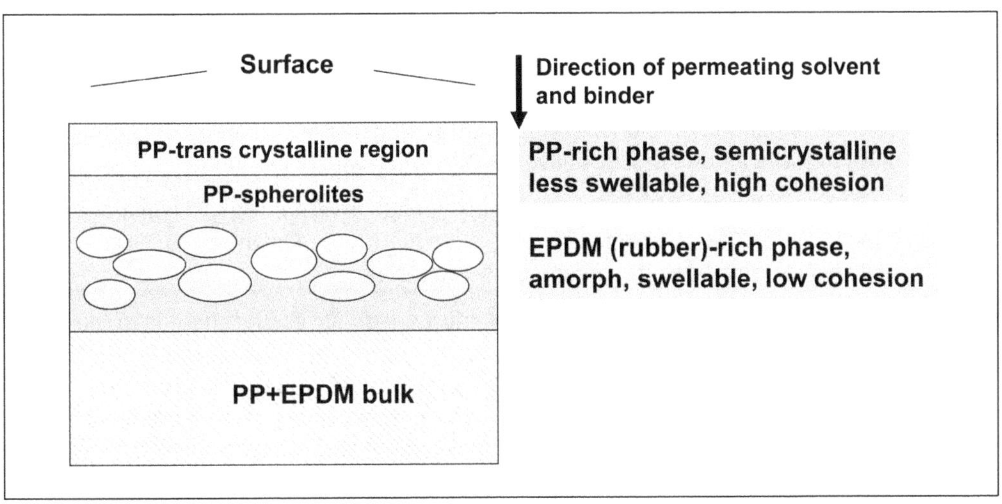

Figure 2.13: TPO, schematic presentation of the shear dependent layer morphology of injection moulded parts

more time to get into its thermo-mechanical state of equilibrium. The resulting local differences in enthalpy are identical with internal stress. The responsible cause, an inhomogeneous orientation of polymer segments can often been made visible by linear polarized light, inducing colourful interference phenomena. Internal stress is often perceived when being relieved suddenly, for example, by crack formation due to solvent attack or by splintering due to mechanical impact. So failure analysis of stress cracking and reduced impact strength of a plastic component involves the investigation of paint and plastic material composition as well as painting and plastic processing.

Mechanical strength

The thermal impact during injection moulding can cause polymer degradation. But there are many other processing related factors, which can have an influence on the thermal mechanical characteristics of plastic parts. Processors have a considerable influence on the plastic formulation and consequently on the technical performance of the substrate to be coated. The plastic raw material intended for injection moulding can be modified with other polymer grades, with recycled material, blends and with aggregates, before being processed. As the development of recycling techniques progresses, as a consequence moulded parts may consist to a certain percentage of recycled material. However the technical characteristics of recycled plastics not always are on the same level as those of not recycled plastics. The thermal mechanical impact during recycling may result in molecular weight degradation. This and the introduction of impurities may induce a loss of quality. Substrates with a higher percentage of recycled material sometimes are not thermally stable, as has been demonstrated by steam jet tests executed on coated parts of TPO.

2.4 Recommendations

On the basis of experiences collected with the manufacturing of coatable moulded plastic parts, principles and recommendations for processors have been worked out, that can be considered as fundamental to achieve success in coating plastic parts [39]. Although these principles are primarily addressed to plastic processors, other participants involved in the production of coated plastic parts should also benefit from them.

At first the plastic part's construction and material composition should be suitable for the coating process. Then the injection mould should be designed with respect to create a proper surface of the component providing the best conditions for coating adhesion. Special attention in tool design and construction should be paid to melt rheology, the uncritical design of weld lines, the avoidance of sink marks, gas inclusions and overheating of the polymer material, as well as enabling a smooth de-moulding. Regarding the process technology of injection moulding the following rules have been proven to be expedient:

- The **tool temperature** has to be adjusted highly enough, in order to avoid condensation of gases, which would affect coating adhesion. A too high rate of cooling causes cold shear on the tool surface, resulting in damage of the material.
- The **melt temperature, shear rate** and **injection speed** have to be adapted to the material. A proper adjustment and control of the melt temperature is of great importance. A too high melt temperature can result in thermal degradation of the material and transport of unwanted ingredients to the surface. Both can be possible causes for later adhesion failures. The polymer melt should keep sufficient retention times over the whole flow path. The shear rate during plasticizing injection therefore has to be adapted to the polymer material. Too high shear forces during injection can cause a flaky surface texture, with negative consequences for the adhesive and cohesive bond strength of the coated part. In

order to achieve a good homogenization, suitable screws with appropriate tip assemblies should be used.

- The **post-injection pressure** should be adjusted properly, at least as high as it takes to obtain a homogeneous melt. An intense polymer homogenization is generally considered as a factor positively influencing adhesion. However a too high post-injection pressure again can deteriorate adhesion.
- **De-moulding** of the injection moulded plastic part should take place without damages and contamination of the surface with de-moulding agents. Tool parts sliding over the freshly moulded plastic can result in surface damages that may cause later adhesion failures. De-moulding agents should be carefully assessed concerning their influence on coating adhesion before coming into use. No silicon-based substances should be used in the entire process of producing coated plastic parts.
- The **ventilation** should be carried out completely. Otherwise air will be trapped in, compressed during injection moulding and thus be heated up strongly. Burn marks will be formed, which can be so strong, that they damage the flow front resulting in bad coating adhesion.
- The **plastic granulate** should be conditioned before coming into use, allowing it to attain a defined humidity. In order to avoid efflorescence or popping it is necessary to perform a well balanced material drying.

2.5 References

[1] PlasticsEurope Market Research Group (PEMRG), The compelling facts about Plastics 2009, publication of PlasticsEurope 2010

[2] G. Stadlbauer, Kunststoffe 95, 10 (2005) 60

[3] R. Mühlhaupt, in Kunststoffe in der Automobilindustrie, VDI-Verlag, 2002

[4] R.A. Ryntz, J.Coat.Tech Research 3, 1 (2006) 3

[5] G. Hilken, Kunststoffe 98, 10 (2008) 91

[6] G. Menges, Kunststoffe 98, 10 (2008) 78

[7] D.W. van Krevelen, Properties of Polymers, Elsevier, Amsterdam Oxford New York Tokio, 1990, Part II

[8] J. Brandrup, E.H. Immergut, Polymer Handbook, 2nd Edition, Wiley, New York, 1975, Chapter III-IV

[9] H. Saechtling, Kunststoff-Taschenbuch, 29. Auflage, Carl Hanser-Verlag, München Wien, (2004) Kapitel 5

[10] H.-G. Elias, Makromoleküle, 6.Auflage, Band 2, Wiley-Verlag Chemie, Weinheim, 2001, Kapitel 13

[11] B. Carlowitz, Tabellarische Übersicht über die Prüfung von Kunststoffen, 6. Auflage, Giesel Verlag für Publizität, Isernhagen, 1992

[12] H. Dominghaus, Die Kunststoffe und ihre Eigenschaften, 5. Auflage, Springer-Verlag, Berlin Heidelberg New York, 1998, Beilage 1

[13] H. Batzer, Polymere Werkstoffe, Band 1, Thieme, Stuttgart New York, 1985, Kapitel 6.1.2

[14] H.G. Elias, Makromoleküle, Band 2, 6. Auflage, Wiley-VCH-Verlag, Weinheim, 2001, Kapitel 10

[15] Allan F.M. Barton, Handbook of Solubility Parameters and other cohesion parameters, CRS Press, Boca Raton, Florida, 1983, p. 153-157

[16] D.M. Koehnen, C.A. Smolders, J.Appl.Polym.Sci. 19 (1975) 1163

[17] T.J. Prater, S.L. Kabeline, J.W. Holubka, R.A. Ryntz, Journal of Coatings Technology 68, 857 (1996) 83

[18] C.H. Hansen, Journal of Paint Technology 39, 505 (1967) 39

[19] W. Hellerich, G. Harsch, S. Haenle, Werkstoff-Führer Kunststoffe, 9. Auflage, Carl Hanser-Verlag, München Wien, 2004

[20] I. Kouleshova, Farbe und Lack 108, 6 (2002) 41

[21] C.A. Harper, E.M. Petrie, Plastics, Materials and Processes, Wiley 2003, New Jersey, p. 520

[22] S. Kirchmeyer, Nachrichten aus der Chemie 10 (2006) 971

[23] M. Osterhold, Vortrag auf der Tagung der Fachgruppe „Anstrichstoffe und Pigmente" der GdCh, Eisenach, 2005

[24] G. Meichsner, Lackeigenschaften messen und steuern, Vincentz Network, Hannover, 2003, Kapitel 5.5

[25] T. Young, Philos. Trans. R. Soc. London 95 (1805) 65

[26] W. A. Zismann, Advan. Chem.Ser. 43 (1964) Chapter 1

[27] D.K. Owens, R.C. Wendt, J. Appl. Polym. Sci. 13 (1969) 1741

[28] C. J. van Oss, M.K. Chaudhury, R. J. Good, Chem. Rev. 188 (1988) 927

[29] F.M. Fowkes, F.L. Riddle Jr., W.E. Pastore, A.A. Weber, Colloid Surf. 43 (1990) 367

[30] F.M. Fowkes, J. Phys. Chem. 67 (1963) 2538

[31] F.M. Fowkes, Ind. Eng. Chem. 56 (1964) 40

[32] S. Paul, Painting of Plastics: New challenges and possibilities, Surface Coatings International Part B: Coatings Transactions, 85, B2, (2002) 79

[33] A. Pfuch, Kunststoffe 97, 3 (2007) 30

[34] M. Rasche, Journal für Oberflächentechnik (JOT) 47, 9 (2007) 46

[35] J. Cremers, Painting of Polypropylen, Seminar: Lackieren von Kunststoffen, Süddeutsches Kunststoffzentrum, Würzburg, 28-29.02.2008

[36] R.A. Ryntz, JCT Coat.Tech. 2, 18 (2005) 30

[37] H.R. Morris, J.F. Turner, B. Munro, R.A. Ryntz, P.J. Treado, Langmuir 15 (1999) 2961

[38] B.D. Pennington, R.A. Ryntz, M.W. Urban, Polymer 40 (1999) 4795

[39] H. Görse, Einflüsse des Spritzgießens auf lackierte Kunststoffbauteile, Seminar: Lackieren von Kunststoffen, Süddeutsches Kunststoffzentrum, Würzburg, 28-29.02.2008

3 *Pre-treatment of plastic surfaces*

The pre-treatment of plastics is often required to transfer the surface into an optimum condition, in which a surface applied coating can possess excellent technical profile without any optical defects. To accomplish this, it is often necessary to perform a preliminary cleaning step, followed by a de-ionization step and, depending on the type of plastic, a final activation procedure. In addition to those measures in advance to the coating process some plastic types need well defined storage under defined atmospheric conditions and/or tempering at moderate temperature. This often includes the following plastic substrates; PMMA, PUR-RIM, SMC and PA (water content).

3.1 *Cleaning*

The high volumes of decorative plastic components used in today's diverse industries have created a large variety of highly specialized coating processes. Some companies produce large amounts of a small variety of big parts, and in general those companies not only coat but manufacture these parts themselves.

Manual cleaning

A well known example of such a plastic part is the automotive bumper, or fascia. The painting process in this case has aspects very much in common with the modern paint lines used for painting automotive car bodies. The processes are very similar and require a high degree of automation, climate control and paint circulation systems to adequately supply paint application robots. In contrast to those companies painting big parts in high volume there are many small paint shops that do handle a large variety of parts, of which each type is produced in small to large amounts. In many cases painting is done manually, without any robots or other automatic equipment. With respect to the diversity of processes, all in all the plastic parts are often pre-treated by blowing ionized air. Sanding is also a frequently used procedure for the elimination of defects and roughens the surface in order to promote paint adhesion. But whether it is appropriate to sand plastic surfaces or not depends heavily on the type of substrate and on the customer's specification for the coating surface, see also Chapter 2.2.5.1.

For example in the serial production of painted SMC-parts for automotive vehicles, sanding is a common procedure. On thermoplastic polymers sanding has a different abrasive effect; in general it causes sanding marks and patterns that are telegraphed to the coating surface. So sanding in general is not a recommended tool for decoratively painted thermoplastic car body parts.

In many paint shops for industrial goods, small exterior and interior car body parts, cleaning is often done manually using dusters that are soaked with water or water/alcohol-mixtures. The fabrics used in dusters differ in quality, and surface cleanliness is often measured

Wilke/Ortmeier: Coatings for Plastics

ISBN: 978-3-86630-859-6

by the level of remaining fibre count per square unit measured. Textile fibres are often a frequent cause for surface irregularities, and while the naked eye may view these as dirt defects, further analysis may be required to confirm defect type and source. When using any type of dusters, synthetic fabrics are preferred over cotton wiping cloths.

When using solvents for cleaning, one has to take care, that the solvents do not attack or swell the substrate, or in some severe cases cause tension cracks. Isopropanol is often used due to its good cleaning characteristics and high polarity.

Power wash

When a large number of components require the optimum in cleaning, one should consider the installation of fully automated in-line washing units. These processes consist of alternating degreasing cleaning zones, followed by stages of de-ionized water rinses to prepare the surface for finishing. In the last zone of this process parts are dried by heating and thereby liberated from adherent water. These processes are considered as the state of the art technology for cleaning large automotive assembly parts. The remaining issues are: incomplete elimination of water from local hollows and niches of three dimensional shaped parts and the subsequent transfer of detergents into the rinsing zones. Determining the electrical conductivity of the last rinsing zone might give information about the contamination with detergents.

Cleaning with carbon dioxide

A possible alternative to power wash is cleaning with carbon dioxide particles, blasted on to the surface either as snow or pellets. The mechanism of cleaning by carbon dioxide snow is described as to be a combination of mechanical impulse transfer, sublimation impulse, thermal stress and the action as solvent [1, 2]. The carbon dioxide particles transmit most of their kinetic energy to the surface and contaminants by impact, where upon they immediately sublime completely to CO_2-gas. The industrial cleaning of ferrous and nonferrous surfaces has made use of this procedure for some time, and plastic parts are just now seeing serial application of this procedure. In order to improve the surface capacity one can work with arrays of jets to maximize cleaning efficiency (Figure 3.1).

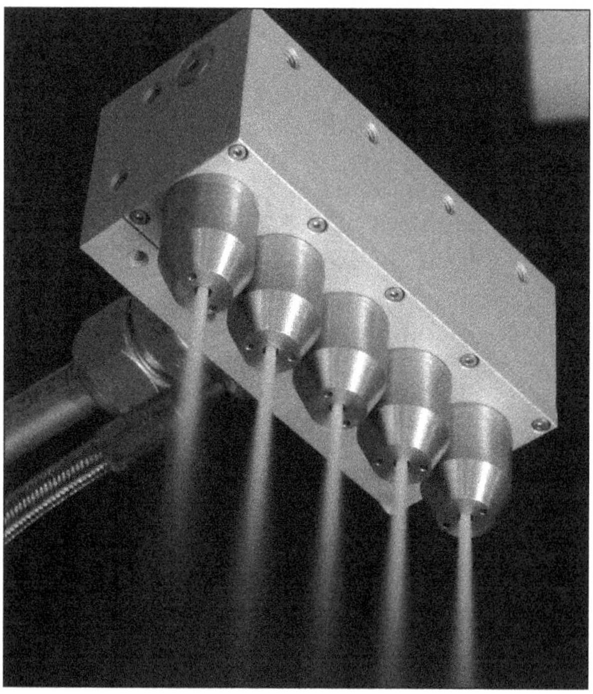

Figure 3.1: Cleaning with carbon dioxide snow by a jet array (photo: acp, Esslingen/Germany [1])

While carbon dioxide snow treats surfaces more smoothly than pellets do, the latter is said to have more cleaning power. Yet it is not clear, whether the cleaning with carbon dioxide is able to achieve those levels of surface quality, in terms of cleanness and smoothness, as well as production speed, that a required in different application areas.

3.2 Activation

3.2.1 Mechanism of polymer surface activation

There are diverse methods of activation to improve paint adhesion on non polar plastics. The primary application of surface activation is for polyolefin's, especially for moulded parts and foils of polypropylene. But also other plastics are activated, for example foils of PVC. The mechanism of action consists of, depending on the applied method, thermal-oxidative, plasma-chemical, wet chemical or photochemical impact (interaction). Although it is not clear yet, how a plastic surface does have to look like in order to achieve good paint adhesion, the mode of activation is probably based on changes in the topography as well as in chemical composition of the surface, resulting from functionalization. The following mechanisms are discussed:

- **Functionalization**: The ability of the surface to establish chemical bonds is increased. Covalent bonds and physical bonds to the coating can be formed [3, 4]. Also an increase of surface tension is taking place, mainly in its polar part, what in this context can be considered as a side effect of functionalization.
- **Increase of surface roughness** :This leads to a stronger interaction with the coating by mechanical interlocking [5].
- **Elimination of so called weak boundary layers (WBL):** The low cohesive integrity of weak boundary layers makes them a possible origin for delamination of the coating system when adhesion tests are performed. In this case the coating is removed from the substrate due to cohesive failure of the plastic substrate. Weak boundary layers in general are generated either by contamination or in the case of polyolefin's by transcrystalline layers of variable thickness [6-8].
- **Interdiffusion of polymers:** This mechanism describes the penetration of polymer chains of the coating into the substrate phase and vice versa. Similar to a mechanical roughening of the substrate's surface, this sort of deeper physico-chemical interlocking should lead to an enhancement of the plastic coating layer composite [9]. The interdiffusion of substrate and coating macromolecules affords a high degree of polymer compatibility.

3.2.2 Analysis of polymer surface activation

There are some powerful analytical methods which can be used for the characterization of polymer surfaces. The element composition of polymer surfaces can be determined by ***X-ray photoelectron spectroscopy*** (XPS) and **e**lectron **s**pin resonance spectroscopy for **c**hemical **a**nalysis (ESCA). In the case of polyethylene and polypropylene there have been

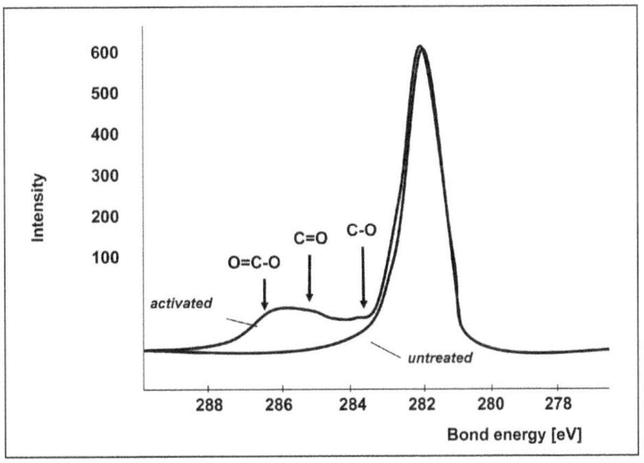

Figure 3.2: XPS of untreated and plasma-treated PE-HD, according [11]

detected hydroxyl-(OH), carbonyl-(C==O) and carboxylic acid groups (COOH), shown in Figure 3.2 [6, 10].

The depth of information with these methods is about 2 to 10 atom layers. XPS can also be used to identify additives on the surface, either from external sources or from the bulk of the polymer, after they have migrated to the surface. Entanglements between substrate and coating polymers can also be examined by XPS and secondary ion mass spectroscopy, as well as by laser scanning confocal fluorescence microscopy [12, 13].

3.2.3 Activation methods

Among experts in the plastic coatings industry, activation methods are often not uniformly classified. At a closer look they show a high degree of diversity. Among the many methods that have been published only those shall be presented, that have gained some significant relevance in the market of industrial plastic coatings technology. Depending on application area and geographic region, they differ a lot in importance and propagation.

A few methods are based on the use of oxidative chemicals, like chromium sulphuric acid or fluorine. A special case is photochemical activation by UV radiation in combination with reactive gases like ozone. In the United States, the preferred method to activate polyolefin's is the application of adhesion promoters based on chlorinated polyolefin's. However in Europe paint lines in general do work with gases in excited or activated state, in most cases oxygen or air. The methods differ from each other in the way, by which those gas excitations have been created. They can be subdivided in methods, which produce cold plasmas by electrically generated discharges at different pressures and those, which produce hot plasma in a gas flame.

3.2.3.1 Flame treatment

As for the coating of automotive plastic parts in the European market, the most common method of activation in Europe is flame treatment. Since the 1930s, flaming has been used to pre-treat plastics on an industrial scale [14]. Main applications today are moulded parts from TPO (PP+EPDM), especially bumpers. Flame treatment systems can easily be integrated into today's highly advanced production lines. They activate plastic surfaces with high reproducibility, even at higher line speeds and they produce little waste. The flame's most active zone has a length of up to 100 mm. This allows the flaming parameters to work with extensive depth, so that parts with strongly varying surface elevations and structures can be pretreated in a well balanced way. Moreover flame treatment offers the option of deburring [15]. In contrast to these processing advantages, there is the fire hazard to be respected, which in particular in the United States is considered as a prohibitive factor.

The gas flames currently in use for plastic pre-treatment can reach temperatures up to 1800 °C. The plastic surface is brought to the highly energetic components of the flame for only 0.01 to 0.1 seconds. The activation effect results from thermal oxidative and plasma chemical processes in the flame. Heat impact, thermal oxidation and plasma reactions lead to morphological and chemical transformations at the polymer surface. Adhering dirt is burned and the polymer surface is oxidized. The mechanical and optical characteristics remain widely constant, as the penetration depth of the flame is only about 20 nm.

Flaming of plastic parts increases its surface energy, as do other oxidative activation methods as well. The surface tension of polypropylene for example is raised from 29 mN/m to approximately 55 mN/m. In general this is considered as a decisive argument for the evaluation of the effectiveness of flame treatment. However, XPS measurements have shown that the oxygen content of the surface correlates better with coating adhesion than the surface

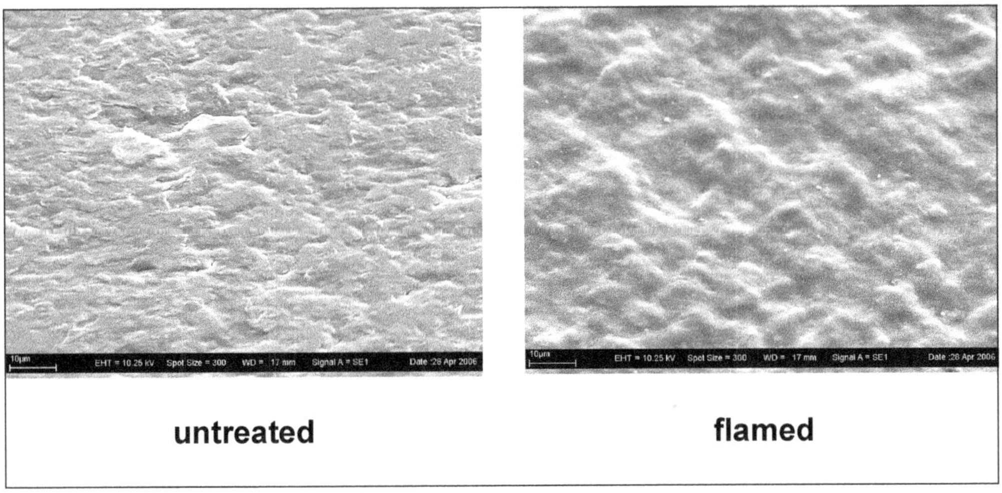

Figure 3.3: SEM of TPO flamed/not flamed [18]

tension [16]. Various functional groups containing oxygen- and nitrogen have been detected by surface analytical methods. These functional groups should increase the bonding capability of the plastic surface to the coating material. Hydroxyl groups that have been created by partial oxidation of the surface, in principal offer the way to a reaction with isocyanate groups of a 2K-coating cross-linker. Physical bonds like dipol-dipol forces, hydrogen bonds and dispersive forces should be enhanced significantly to increase coating adhesion. It has been observed also, that the concentration of hydroxyl groups of the surface goes proportional with the number of flame treatment cycles [17].

Side effects of flame treatment are: changes of the surface topography/morphology, and an erosion of the substrate material. Modification of the surface topography by flaming, in the case of TPO can be visualized well by electron microscopy shown in Figure 3.3.

Depending on the moulded part's geometry and on the technical circumstances in the paint shop facility, flame treatment can either be executed by moving the part or the flame.

Industrial plastic coating processes today mostly use automatic flaming robots equipped with burners producing wide flame shapes. A homogeneous activation of the plastic surface then affords movement of the flame along well programmed paths. A flame treatment system consists of an adequate burner and a control unit. Flaming efficiency depends on:

- Type of burner
- Kind of gas used
- Mixing ratio gas/air
- Gas flow rate
- Distance of the burner to the surface
- Speed of the flame/object

In many cases the type of burner is dictated by part geometry and paint shop lay out, whereas the gas flow rate is depending upon design and capacity of the burner. There is a certain degree of freedom concerning the choice of flame gases and mixing ratios. Common flame gases are methane, propane and butane. In mixtures of propane and butane, one should respect the tendencies of gas separation in the gas storage unit. When butane is used, it is known to accumulate in the gas storage unit over time so care must be taken to ensure consistency.

The mixing ratios gas/air can be calculated easily, on the basis of the reaction equations for burning processes and the oxygen content of the air. Burning of one mole of propane needs five moles of oxygen, which in the case of methane only is needed in two moles (Equations 3.1. and 3.2).

Equation 3.1: $C_3H_8 + 5O_2 \rightarrow 3CO_2 + 4H_2O$

Equation 3.2: $CH_4 + 2O_2 \rightarrow CO_2 + 2H_2O$

Based on the fact that air contains 20 to 95 Vol.-% oxygen, the stoichiometric combustion of propane with air affords a mixing ratio propane/air of about 1 : 24 (100/20.95 · 5.0=23.86). If methane is to be used instead of propane, the correct calculation results in a mixing ratio of methane/air of 1 : 10. In reality a stoichiometric oxidation often is not the goal, but flaming parameters are driven in a way, that a flame treatment with excess oxygen, often with 10 %, or as the opposite, reductive flaming is executed [16].

The usual speed of the burner unit ranges between 5 and 25 m/min, distances of burner to object between 4 and 20 cm.

It has been found that the effect of activation does not correlate in a linear way with the intensity of flaming. The oxygen content of the surface increases with growing burner speed up to a maximum, then it declines again. The same applies to the distance of the burner to the substrate. In order to achieve a maximum concentration of oxygen on the surface, the combination of speed and distance is decisive. For a given design of moulded parts and burner unit, it should be possible to find an optimum of flaming parameters that produce a maximum amount of oxygen on the surface shown in Figure 3.4 [16]. Similar connections have been described for burner speed/distance and surface tension [15].

As for polypropylene it has been found that coating adhesion correlates with the oxygen content of the surface. Coating adhesion was ok above an oxygen value of about 12 % at the

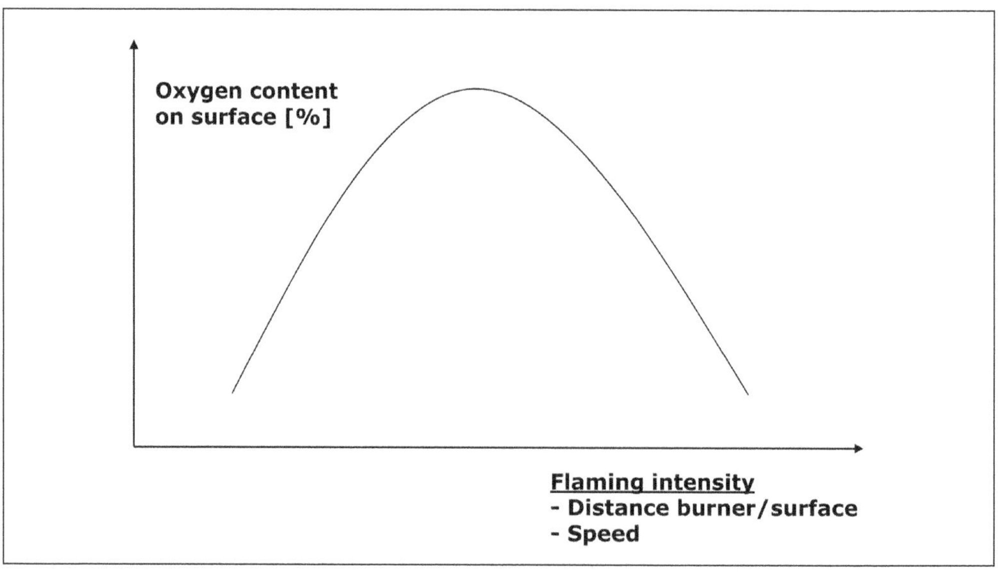

Figure 3.4: Schematic diagram of the surface oxygen content depending on the intensity of flame treatment

surface. However, no correlation between coating adhesion and contact angle after flame treatment could be established. Another interesting observation was that the oxygen concentration inversely correlated with the surface temperature. Concerning coating adhesion, the optimum temperature range was between 50 and 80 °C [16].

3.2.3.2 *Principles of plasma treatment*

A plasma stream consists of a mixture of positively charged ions, electrons, neutral particles and highly energetic molecular fragments [19]. It is generated using ignited gases with the addition of electric tension or electromagnetic waves [11]. The current in an electrically charged gas is called a discharge [20]. In this sense plasma is an extreme kind of a discharge. Gas flames are thermal plasmas and that is why flame treatment can be considered as a special case of plasma based activation. On the other hand discharges generated by microwaves or electrical energy produce much less thermal impact. Their temperature, perceived from the outside, hardly exceeds 100 °C. Procedures that are derived from electrical discharges cover plasma treatment methods by reduced pressure and ambient pressure, including corona methods.

A plasma treatment only modifies the top ten nanometers of a polymer surface. Furthermore the polymer surface is not degenerated significantly [17].

Discharges are often associated with luminescence, which is a result of the recombination of energy rich charged particles that emit their energy as light. Another side effect can be observed with corona treatment: Shining channels of discharges between electrodes that have the character of current stitches, which are called plasma filaments. Plasmas with such kind of current filaments are not potential free. Apart from this special case, plasmas are, despite their numerous electrical charges, electrically neutral to the exterior.

The type of discharges used in plasma technology is called independent. The kinetic energy of the charged particles is so high, that they produce new charges by collision in an amount, that a complete substitution of recombined and therefore neutralized ions and electrons is secured. A sufficient high current density will lead to a stable and enduring plasma state

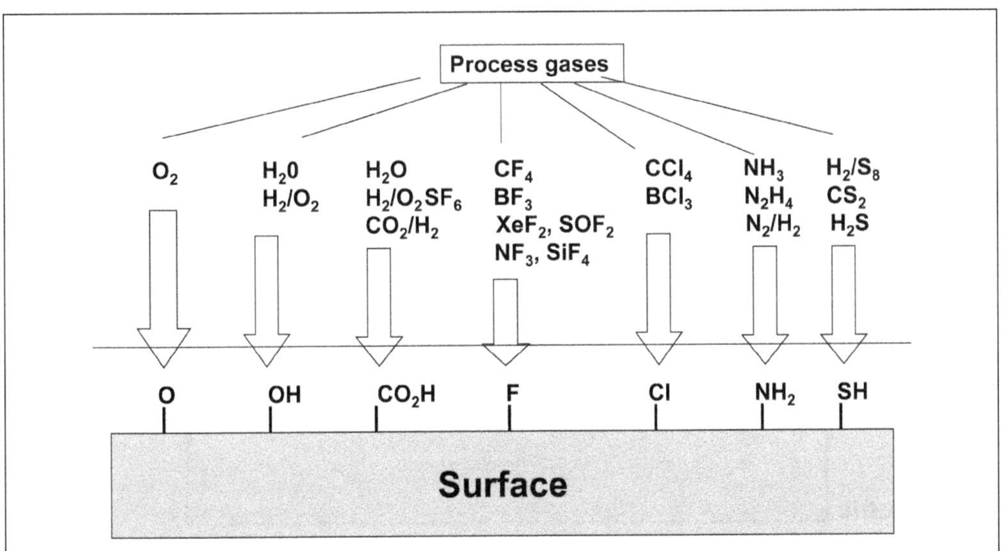

Figure 3.5: Plasma process gases, according to [11]

and the already mentioned luminescence light. Due to its energy rich particles plasmas are chemically reactive. The common principle of all plasma based activation methods is the chemical modification of polymer surfaces by ionic and radical compounds of the process gas, mostly being air.

The process gas can be chosen in a quite flexible way. Consequently plasma activation offers a broad variability of chemical modifications (Figure 3.5), which opens the way to tailor suite the functionalization process.

For example, if the process gas contains oxygen by using air, the resulting plasma will oxidize polymer surfaces of plastics or coatings [21]. However, if pure nitrogen is used, a functionalization with amino groups will become possible.

Plasma methods differ from each other not only by the kind of gas being used, but also by the working pressure, the spatial reach and the homogeneity of their plasmas as well as by their mode of application, either in a continuous or discontinuous way. Consequently there are plasma methods more appropriate for moulded parts and those that are preferred for foils, which maximize the specific functionalization of polymer surfaces, with hetero atoms (O, S, and N).

3.2.3.3 Plasma treatment at reduced pressure

Treatment of plastic surfaces with plasma at reduced pressure is carried out in a vacuum chamber, equipped with electrodes, measurement and control units, as well as with a gas inlet and outlet. Process gases used today for the activation of polymer surfaces include air, oxygen, nitrogen and argon, whereby partial mixtures are also in use. At a pressure of 10 to 200 Pa (0.1 to 2 mbar) the gas is discharged and the plasma ignited by alternating voltage or radio-/microwaves, see Figure 3.6.

The so created electrons, ions, neutral particles, radicals and photons react with the polymer surface. The resulting newly created radicals can further react with molecules of the process gas, by which they are transformed into stable functional groups. After stopping

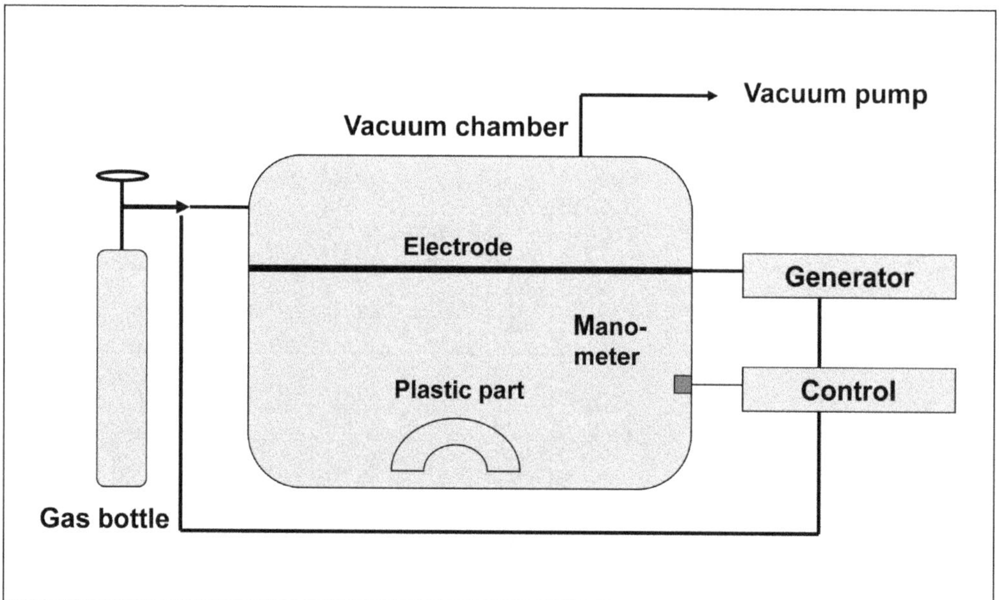

Figure 3.6: Reduced pressure plasma unit[22]

the energy supply, the plasma will collapse, allowing the gas to return to its neutral initial state (switchable plasma). Although the ions of the plasma can have temperatures from 100 to 1000 Kelvin, and the electrons even 10^4 to 10^5 Kelvin, the overall temperature of the plasma normally ranges 300 to 400 Kelvin. The general assumption is that the plasma is not in a thermodynamic equilibrium, in contrast to hot plasma, which for example may be generated by nuclear fusion or a thermal flame. Therefore plastics are stressed thermally much less by reduced plasma treatment than by flaming. Further advantages of plasma treatment by reduced pressure are the homogeneity of the plasma, the uniformity of activation even at zones of three dimensional parts that are less accessible, as well as the high level of control and reproducibility of the process. From the point of view of industrial safety and environmental protection, other advantages are: The isolation of the gas chamber and the practicability to remove all gaseous substances by vacuum. Another benefit of reduced pressure plasma technology is beyond the scope of activation but worth discussion: The feasibility to use this process for the coating of plastics with diverse substances from the gas phase (chemical vapour deposition: CVD). An example here is gas phase polymerization.

On the other hand, a disadvantage of low pressure plasma treatment is that it has to be carried out in a discontinuous way. The batch character and the obligation to vacuum technology account for the comparable high process costs of this kind of activation method. Reduced pressure plasma treatment is especially suited for the activation of bulk goods and small parts made of plastic. However, plasma chambers can be incorporated as well into production lines for automotive parts and therefore are also operated **in-line**.

3.2.3.4 Plasma treatment at ambient pressure

Open plasma can be produced by application of a sufficient high direct or alternating voltage between two electrodes. Two process variants can be distinguished, depending on the way the plasma is guided to the part that has to be treated. One can either create direct discharges, bringing the resulting plasma directly to the surface, or chose for indirect discharges, in which a detour is used.

Modern developments use gases and aerosols for specific chemical functionalization. Plasma treatment with mixtures of nitrogen and hydrogen allows the selective formation of amino groups on the surface [21].

Direct corona treatment

The activation of polymer foils by barrier discharges is known as direct corona treatment. This can be considered as a special kind of plasma treatment at ambient pressure, using two asymmetrical electrodes, an electrical insulator and the process gas. The two electrodes have a length fitting to the width of the foil and form a gap of approximately 1 to 2 mm, in which the foil is exposed to electrical discharges (Figure 3.7).

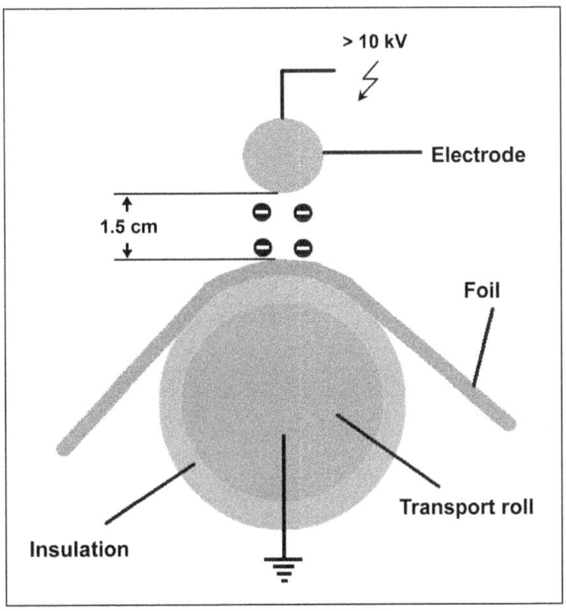

Figure 3.7: Direct Corona treatment [18]

One of the two electrodes is the transport roll and equipped with a dielectricum, by which it is supplied with an insulation effect. By application an alternate voltage of typically 15 kV and a frequency of 30 kHz are used and the gas in the space between the two electrodes becomes discharged. The gas mostly consists of air, which flows around the electrodes, in order to cool them and remove the by-product ozone. Depending on the level of voltage, gas discharge channels and the typical corona are formed. Speaking in a figurative way a corona can be described as a "wind of ions". Although this activation method itself is named after it, the corona is not the most active part of the plasma. The plasma filaments (gas discharge channels) are of much higher energy in contrast to the rest of the plasma, as the filaments are not potential free and should not be touched. The energy resulting from a potential of some hundred volts is high enough to break chemical bonds and produce radicals at the polymer surface. Similar to flame treatment and low pressure plasma treatment, functional groups are formed, the kind being dependent on the type of used gas. The heterogeneous distribution of energy and the high energy of the plasma filaments contain the risk of local overheating. The barrier effect of the transport roll helps to avoid combustions which otherwise would be the result of a too high local current density and electrical breakdowns. Also it leads to a homogenization of the electrical field on the surface of the foil. The direct corona treatment is mainly applied for the activation of foils when polyolefin's and PVC are the chosen materials. These can be activated in a width of 10 m with a line speed of 1000 m/min. The ozone being produced in operation of a corona plant has to be removed by vacuum. Normally air is used as the gas in this process. However, processes using pure nitrogen or nitrogen/hydrogen mixtures have been developed. Hereby the surface can be equipped with amino groups that can be connected with coating systems via covalent bonding and therefore promoting adhesion. In order to deposit specific coating layers also aerosols can be applied instead of pure gases. The advantages of direct corona treatment are high transport speed and efficiency. Due to geometrical limitations - pre-treatment in a gap of 1 mm - the applications are restricted to webs/sheet products.

Indirect corona treatment

This activation method, also known as spray corona method, has been developed to extend the application to three-dimensional plastic parts. The discharges are produced in a housing

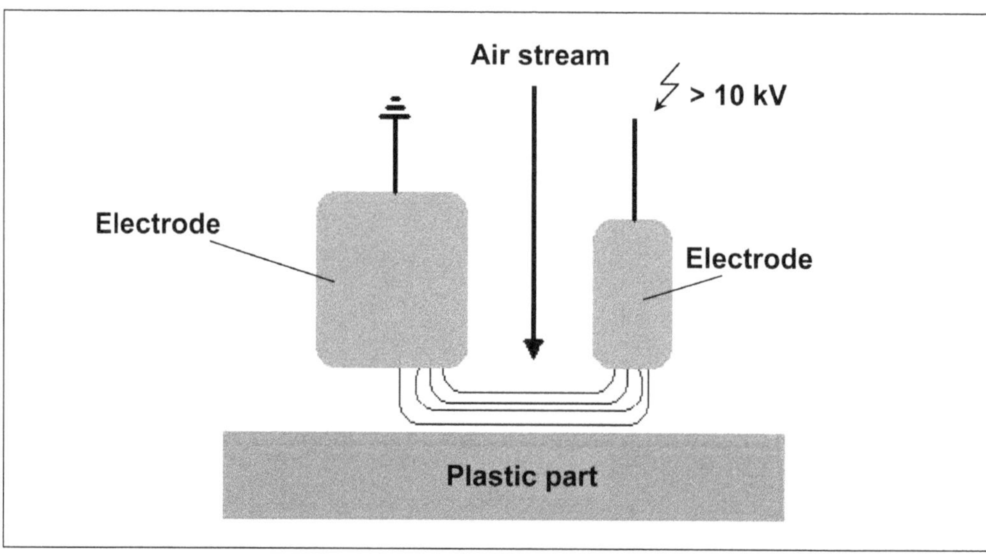

Figure 3.8: Indirect corona treatment[18]

whereby the plasma filaments are guided through an outlet of that housing by a stream of air (Figure 3.8) The central temperature of the plasma does not exceed 50 °C under standard conditions, but it is not potential free and can therefore not be touched without danger [23].

Plasma jet treatment

This sort of plasma treatment at ambient pressure is another indirect activation method. Similar to the indirect corona treatment the discharge system consists of a housing, whereby the plasma is created between two electrodes, then separated from the plasma filaments and guided to the piece by pressured air shown in Figure 3.9 [18].

The speed of the created plasma jet can be 120 to 300 m/s. In contrast to the indirect corona treatment, the obtained plasma is not under potential and has a temperature of about 60 to 100 °C. The treatment speed is comparable with that of direct corona methods. The shape of the nozzle determines the shape of the treated surface area. Besides single round shaped nozzles there are an array of various nozzle configurations for treating broader surface area activation simultaneously. Current work is focused on the homogeneity of the treated surface. Examples for application are profiles, closure caps and housings of cellular phones [24].

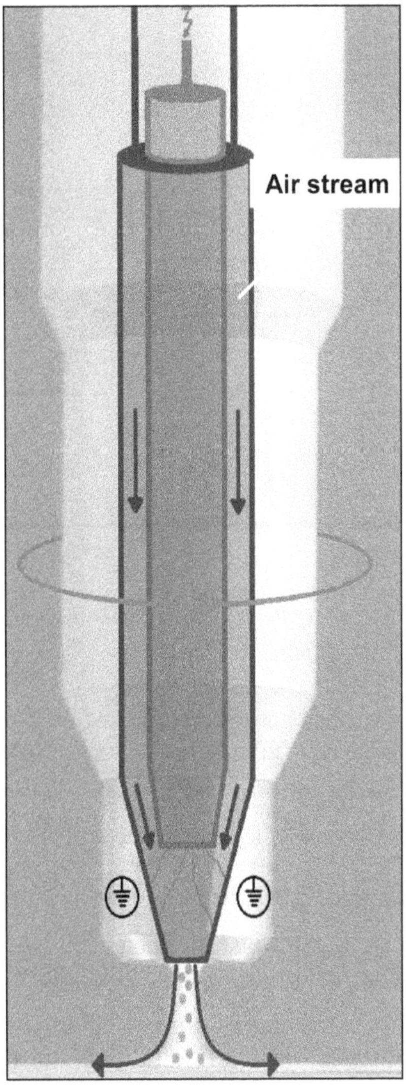

Figure 3.9: Plasma-jet treatment

Benefits of the plasma jet treatment certainly result from a combination of the advantages of the before mentioned methods. It can be applied with high in-line speeds continuously at ambient pressure. In contrast to direct corona treatment it not only represents a continuously working plasma-based activation method, but also is appropriate for three dimensional parts, under potential free conditions. The latter has more in common with the low pressure plasma methods, which however, are executed in a discontinuous way.

3.2.3.5 Fluorination

This activation method is based on the exothermic reaction of fluorine gas with hydrocarbons. This reaction includes the cleavage of carbon hydrogen bonds, the formation of hydrogen fluoride and of surface near carbon-fluorine bonds (Equation 3.3). The latter being more polar than the carbon hydrogen bonds, which results from the high electro negativity of fluorine.

Equation 3.3: $-CH_2- + 2F_2 \rightarrow -CF_2- + 2\,HF$

Technical fluorination of polyolefin surfaces does not lead to a complete conversion of all carbon hydrogen groups. This is the best achievable result, as per fluorination would lead to a hard to wet "Teflon"-like surface. Fluorination can lead to an increase of the polar part

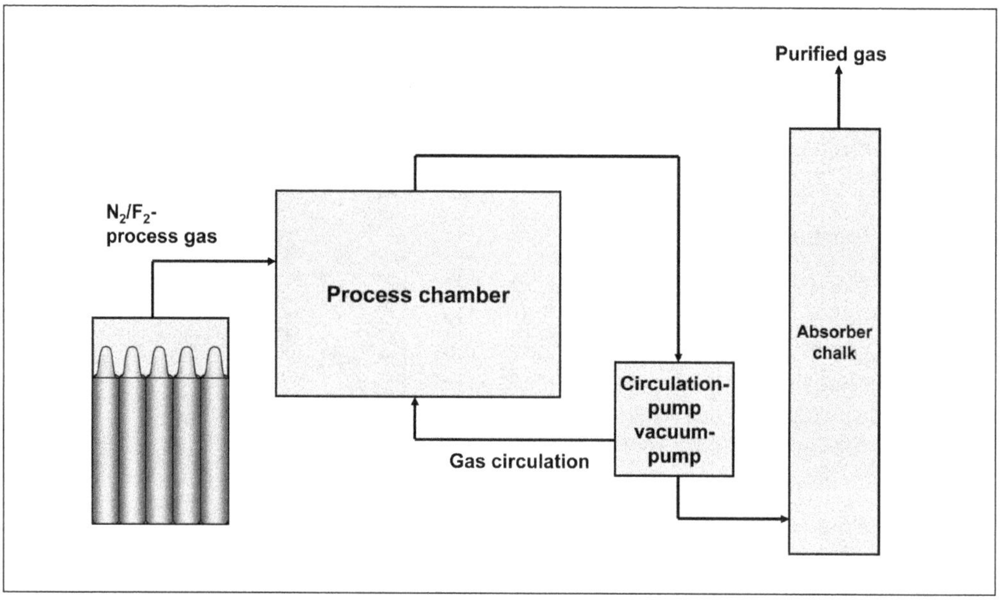

Figure 3.10: Plant for discontinuous fluorination, according to [26]

of surface energy from near 0 mN/m up to more than 10 mN/m, the total surface energy rising from 28 mN/m to 66 mN/m [22, 25]. Fluorination is executed mainly in a discontinuous way, as a batch process (Figure 3.10).

A fluorination chamber has a volume of up to 20 m^3 and works at temperatures of 55 to 70 °C [26]. While small pieces are preferably treated in cylinder shaped reactors, for larger moulded parts, rectangular reactors are often used.

Although the discontinuous fluorination is of higher (relevance) importance, the continuous fluorination of extrudates like sheets and foils is also possible [27].

Due to the high exothermic fluorination reaction, the gaseous fluorine has to be diluted with nitrogen, leading to a fluorine content of 0.1 to 10 % [22]. Evacuation of the reaction chamber down to residual pressure of 0.1 to 1.0 mbar is followed by filling with the reaction gas to a pressure of 10 to 900 mbar. The parts are typically contacted evenly to the gas for about 15 to 120 seconds. Inner surfaces, for example with bottles or textiles can be pre-treated effectively without any problems. The hydrogen of C-H-bonds is not only substituted by fluorine but also by oxygen, as could be demonstrated by ESCA. The cause for this side reaction is the residual oxygen in the process gas [27]. In order to terminate the fluorination process, the gas is simply pumped out of the chamber [26]. The hydrogen fluoride, being formed as a by product, has to be collected in an absorber system, containing chalk. This is subsequently transformed into calcium fluoride, a compound that also is formed by reaction of chalk with excess fluorine gas (Equations 3.4 and 3.5).

Equation 3.4: $2HF + CaCO_3 \rightarrow CaF_2 + H_2O + CO_2$

Equation 3.5: $F_2 + CaCO_3 \rightarrow CaF_2 + CO_2 + 1/2\ O_2$

After rinsing the reactor, the parts can be taken out. In long term trials it could be shown, that the surface tension did not decline after aging for 21 months. Apparently the C-F-bonds being established by fluorination are of high durability, which might be caused by the high penetration depth of several atom layers. Another result that has potentially a promoting effect on adhesion is the increase of roughness [26]. A precondition for impeccable fluorination is a clean surface, free from any kind of contamination. Application areas of the fluorination are various parcels like automotive interior parts and tool grips. Apart from the importance of increasing the wet ability for coatings and adhesives, fluorination serves for improvement of other material characteristics, like the reduction of gas diffusion and glide friction.

3.2.3.6 Other methods

UV radiation

By this method plastic surfaces are exposed to UV light of 135 to 180 nm in an ozone containing atmosphere. Thereby polymer radicals are formed, that further react with ozone. The depth of treatment is about ten nanometers; the speed varies from one to four minutes. Only flat parts and no three dimensional complex shaped objects can be treated in this way. During the process the temperature increases to 40 to 45 °C. The surfaces have to be free of contaminations prior to further treatment. The activation by UV radiation and ozone has been applied for instance in the industrial manufacturing of shoes [21].

Wet chemical pre-treatment

For many years, polyolefin's have been activated by processes using wet oxidative chemicals. This kind of pre-treatment involves the contact of plastic surfaces with strongly oxidative mineralic acids for a certain period of time. Examples of those combinations are mixtures of sulphuric acids and sodium bichromate, manganate or chlorate. This kind of activation is applied in the industrial pre-treatment of squeegees for window wiper blades.

Adhesion promoters for polyolefin's

The application of adhesion promoters in thin layers is a well known and established technology in North America to improve coating adhesion on polyolefin surfaces. This method helps to avoid open flames, and compared to plasma technologies, it is considered to be less capital intensive and easier to be controlled. Plants using adhesion promoters instead of one of the before described activation methods are not obliged to a certain shape or geometry of parts and do not need capital extensive equipment, but just standard painting application equipment. Adhesion promoters designed with chlorinated polyolefin's (CPO-primers) can get into interaction with polyolefin surfaces in a way that a strong adhesive composite of substrate and coating polymer is established. The adhesion mechanism probably is based on a combination of *van der Waals*-forces, molecular entanglements respectively interdiffusion of polymer chains. *Prater* et al. have investigated the diffusion of CPO-based adhesion promoters by secondary ion mass spectroscopy and fluorescence microscopy [28].

However the low solid content and high level of chlororganic compounds of CPO-primers are of disadvantage. Water dispersability can be achieved by modification of the adhesion promoter resin with maleic anhydride groups, subsequent hydrolysis and neutralization of the carboxylic acid groups [29].

3.3 References

[1] C. Lengerer, Kunststoffteile mit CO_2 effizient reinigen, Seminar: Lackieren von Kunststoffen, Süddeutsches Kunststoffzentrum, Würzburg, 28- 29.02.2008

[2] A. Michalske, Journal für Oberflächentechnik 47, 4 (2007), 76

[3] J. F. Friedrich, S. Geng, W. Unger, A. Lippitz, J. Erdmann, H. V. Gorster, C. Wöll, A. Schertel, K. Bierbaum, Surf. Coating Technol. 664 (1995), 74-75

[4] E. Papirer, D. Y Wu, J. Schultz, J. Adhesion Sci. Technol. 7 (1993), 343

[5] W. Brockmann, Adhäsion 22 (1978), 6

[6] K. Jänichen, J.Adh.Science Tech. 17, 12 (2003), 1635

[7] J. J. Bikerman, D. W. Marshall, J. Appl. Polym. Sci. 7 (1963), 103

[8] H. Schonhorn, F. W. Ryan, J. Appl. Polym. Sci. 18 (1974), 235

[9] L. H. Sharpe, Adhesion 67 (1998) 277

[10] D. Briggs, D. M. Brewis, M. D. Konieczko, J.Mater.Sci. 14 (1974), 1344

[11] G. Ellinghorst, Plasmen und Plasmatechnik, Bremen, 2007

[12] Y. Ma, J. P. S. Farinha, A. Winnik, P. V. Yaneff, R. A. Ryntz, Macromolecules 37, 17 (2004), 6544

[13] Y. Ma, J. P. S. Farinha, A. Winnik, P. V. Yaneff, R. A. Ryntz, J.Coat.Tech. 2, 5 (2005), 407

[14] M. Hill, Gasflammenbehandlung von Kunststoffen, Industrie-Lacktagung, VdL-VILF-Vincentz, 1995

[15] W. Eckert, Deburring/deflashing and surface treatment of plastic surfaces by flaming, European Coatings Conference, Modern Coatings for Plastic Substrates III, Vincentz Network, Berlin, 15-16.03.2007

[16] J. Cremers, Painting of Polypropylene, Seminar: Lackieren von Kunststoffen, Süddeutsches Kunststoffzentrum, Würzburg, 28.-29.02.2008

[17] S. Paul, Painting of Plastics: New challenges and possibilities, Surface Coatings International Part B: Coatings Transactions, 85, B2, (2002) 79

[18] D. Cilbuka, Diplomarbeit, Fachhochschule Esslingen, 2006

[19] BMBF, Referat für Öffentlichkeitsarbeit, Bonn, 2000

[20] P. Dobrinski, G. Krakau, A. Vogel, Physik für Ingenieure, Teubner, Stuttgart, 1984, Kapitel 3.2.6

[21] K. W. Gerstenberg, Pretreatment of Plastic Parts, European Coatings Conference, Modern Coatings for Plastic Substrates III, Vincentz Network, Berlin, 15.-16.03.2007

[22] G. Meichsner, T. G. Mezger, J. Schneider, Lackeigenschaften messen und steuern, Vincentz Network, Hannover, (2003) Chapter 6

[23] Tigres GmbH, Publikation 44: Kunststoffteile vorbehandeln mit Plasma ohne Vakuum

[24] S. Fischer, Die richtige Aktivierung, Seminar: Lackieren von Kunststoffen, Süddeutsches Kunststoffzentrum, Würzburg, 28.-29.02.2008

[25] A. Pfuch, Kunststoffe 97, 3 (2007) 30

[26] B. Möller, Kunststoffe 89, 12 (1999) 41

[27] T. Zeiler, F. Achereiner, Kunststoffe 97, 7 (2007) 30

[28] T. J. Prater, S. L. Kabeline, J. W. Holubka, R. A. Ryntz, Journal of Coatings Technology 68, 857 (1996), 83

[29] M. D. Foster, S. F. Thames, Journal of Coatings Technology 71, 889, (1999) 91

4 Coating materials

4.1 Industrial plastic coating

Plastic parts are coated for functional and decorative reasons. With these considerations in mind, in combination with the type of plastic used, questions of economy and environmental aspects, various material concepts are available (Figure 4.1).

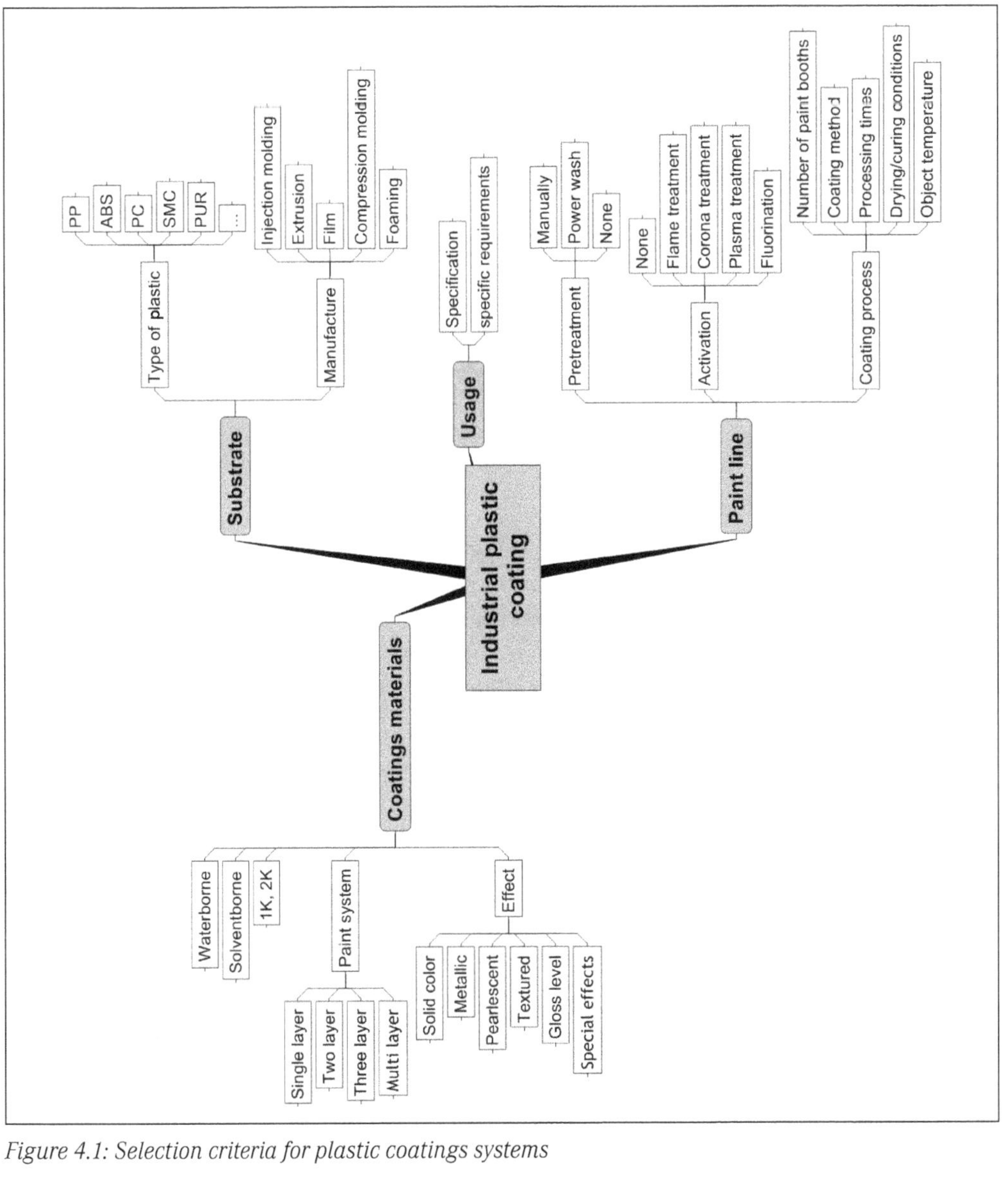

Figure 4.1: Selection criteria for plastic coatings systems

Wilke/Ortmeier: Coatings for Plastics

ISBN: 978-3-86630-859-6

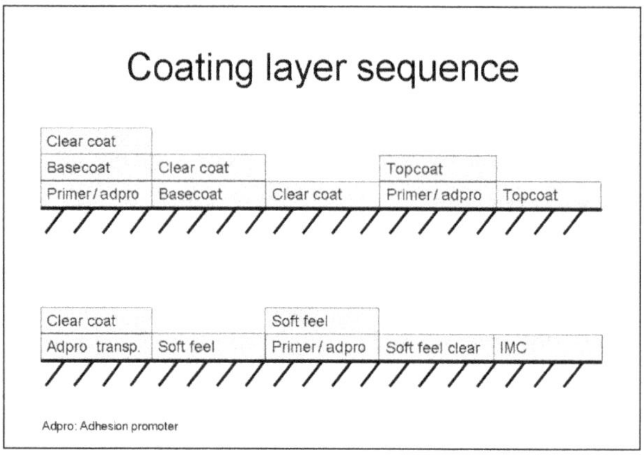

Figure 4.2: Options for layer sequences/layer structures

Planning of the coating system usually begins with the question of end use: cell phone shell, automotive interior or exterior part, television cabinet to name just a few.

Apart from these considerations different principal possibilities of layer sequences exist shown in Figure 4.2.

Most common practices used for many years, especially in the automotive sector, is the so-called three layer system consisting of a primer (first layer), solvent or waterborne, or a solventborne adhesion promoter (adpro) in combination with solvent- or waterborne basecoat and solventborne clear coat. Usage of waterborne clear coats is at present limited to a few exceptions.

The relevance of the different technologies varies by region. In Germany and its European neighbourhood water-based primers and basecoats are widely used whereas in France, Spain, Italy and some eastern countries solvent-based systems are still preferred. In the US adhesion promoters containing 95 % organic solvents, cured at high bake temperatures, are used in order to avoid flame treatment, a predominant process used in Europe for painting PP (polypropylene) blends.

After many years of research to find more economical coatings solutions by reducing the film thickness of the individual coatings layers reached its technological limits. Work was then started to omit entire paint layers. These days the two layer system without primer is very popular. As a consequence the basecoat must ensure, besides colour and effect, the important function of adhesion on the different substrates.

The **direct to plastic** coating of mass coloured plastics (moulded in colour, MIC) or plastics with special designs with clear coats has only been done in larger commercial scale since the end of the 90s of the 20th century. The requirements for clear coats for this application are discussed in Chapter 4.6.

Topcoats can be found as solid shade paints, as effect (metallic, pearlescent) or textured materials, in one or two layers, as solvent as well as waterborne systems. A special type of topcoats is **soft feel coatings**, where today more than 90 % are comprised of water-based formulations, which give a pleasant touch or haptic feel to the thereby coated articles.

For **in mould coating** (IMC), the coating is applied, as the name indicates, directly into the mould. This can be done prior to the moulding of the part itself by painting the open cavity or afterwards by injecting the IMC paint under high pressure into the slightly opened mould.

A special coating process is the **painted film technology** where in a first step a foil is painted, then formed finally becoming the painted part by back moulding, back foaming, adhesive bonding or laminating.

The individual coatings materials will be described in greater detail in the sections later on.

4.2 *Adhesion promoters*

Generally spoken this product group refers to materials which normally are applied at low film builds, are often colourless and for the most part have very low solid content. Adhesion promoters (adpros) are usually over-coated after short flash-off without intermediate drying.

Adpros are most common for the coating of thermoplastic polyolefin (TPO). They are made from highly diluted solutions of chlorinated polyolefin (CPO) in aromatic hydrocarbons. The adpros develop their full effect when in the subsequent process steps temperatures above 100 °C arise. At this temperature level the CPO fuses to the TPO substrate establishing an optimum compound.

Such systems are hardly used any more in Germany because of their high content of solvents, often reaching 90 % and higher. Also due to the fact that in general lower curing temperatures are used, activation, e.g. by oxidizing gas flame, is more the standard in Europe, and the adhesion promotion becomes an integrated function of the primer system. In the US however, so-called high bake systems with curing temperatures of 120 °C are used allowing for a more widespread use of adpros, whereas flame treatment is often not considered. Many years of research to make hydro adpros perform like solventborne adpros in the application process have been pursued, and thus far it has been difficult to emulate the adhesion properties of the solventborne types in today's marketplace. Other efforts have the aim to develop waterborne, halogen-free alternatives to be able to offer products in a more sustainable delivery form. First raw materials for this new type of adpros are on the market since recently and now have to prove their ability to compete.

Adhesion promoters are used in other applications as well: under clear coats for carbon fibre reinforced polymers (CFRP), under clear coats on moulded in colour parts (MIC), to improve foam adhesion on plastic frames e.g. for the production of instrument panels for cars.

4.3 *Primers and fillers*

Primer denotes the first layer of a multi layer system. As a rule its film build is <20 µm. The exception would be coloured or white primers, which often require film builds up to 30 µm. Fillers can also serve as a first layer, and can be used in conjunction with primers to meet other application or performance properties. With applied layers of >30 µm, occasionally up to 100 µm, fillers are used to level out surface unevenness and defects like striae, pores, blow holes or flow lines. An overview of the functions or primers and fillers is shown in Figure 4.3.

Table 4.1: Guideline formulation for a waterborne adhesion promoter [1]

Item	Raw material	wt.-%
01	CP-Dispersion	29.1
02	Polyurethane thickener	0.9
03	Defoamer	0.01
04	Demineralized water	69.99
	Total	100.0

Item 01: Eastman CP 310W, CP 347W oder CP 349W
Item 02: Henkel DSX 1514
Item 03: Byk 022
Note: Predilute DSX 1514 in partly or total quantity of water and add while stirring

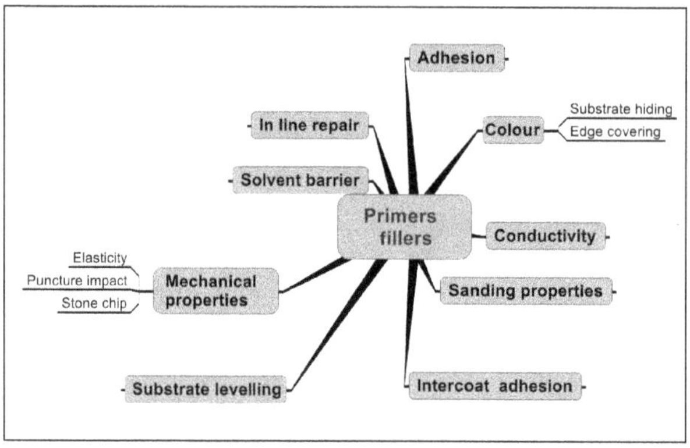

Figure 4.3: Major properties of primers and fillers

The most important feature of primers is the adhesion to the substrate and the intercoat adhesion to the succeeding layer. The primer must level out surface unevenness, be easy to sand; it has to hide the often black colour of the plastic substrate, and support hiding of the overlying layers, while providing impact and stone chip resistance of the entire compound and subsequent coating system. The primer can have a barrier effect against solvents from the subsequent layers to protect solvent sensitive plastics. In many cases primers are formulated with conductive pigments for the still growing electrostatic application of basecoat and clear coat.

The most common primer colour is light to dark grey. In special cases they are pigmented in analogue colours for basecoats or topcoats with limited hiding power. The most important characteristics of typical 1K and 2K primers are shown in Table 4.2.

Primers can be 1K or 2K, solventborne or waterborne. The advantage of 1K materials is their good workability. 2K systems often have better performance, notably in terms of adhesion and in particular under humid conditions (humidity chamber, weathering tests). Another special application is primers for IMC (in mould coating). The application of the primer layer is transferred to the injection mould or pressing tool. Challenging is that the coating has to be solvent-free. Otherwise surface defects, pin holes or blisters can occur by solvents or water entrapped in the mould.

The resin base for primers and fillers for plastic substrates are flexible, water or solventborne polyester, polyurethane-polyesters or polyurethanes. To control drying or sanding, acrylics can be combined (thermoplastic or cross-linkable). In this case special care has to be taken for the mechanical properties.

Table 4.2: Typical properties of primers

	Waterborne		Solventborne	
Colour	light, medium, dark grey black, transparent, colourless	white, colours	light, medium, dark grey black, transparent, colourless	white, colours
Solids [wt.-%]	approx. 30 to 35	approx. 35 to 45	approx. 20 to 35	approx. 25 to 40
Binder-to-pigment ratio	approx. 1 : 2	approx. 1 : 2.5	approx. 1 : 2	approx. 1 : 2.5
Density [g/cm³]	approx. 0.9 to 1.1	approx. 1.2	approx. 1.1	approx. 1.1
Flow time [s] DIN4 cup	18 to 30	25 to 40	18 to 30	18 to 30
Drying/curing	typical 15 to 30 minutes 80 °C		some minutes @ RT up to ≥30 minutes ≥80 °C	
Film build [µm]	10 to 15	15 to 30	10 to 15	15 to 30
Dry film conductivity [kΩ]	<10 up to >100		<10 up to >100	

Table 4.3: Guideline formulation of a 2K solventborne primer[2]

Item	Raw material **Component 1**	wt.-%
01	Polyester, OH-functional	9.4
02	Thermoplastic acrylate, 20 % in ethyl acetate	23.5
03	Diacetone alcohol	19,8
04	1-methoxypropylacetate-2	10.0
05	Titanium dioxide rutile type	11.6
06	Talc	4.6
07	Calcium carbonate	7.0
08	Bentone 38, 10 % intermediate	4.8
	Component 2	
09	Aromatic polyisocyanate	9.3
	Total	100.00

Item 01: Desmophen 1200, Bayer
Item 02: Paraloid B48N or Paraloid B99, Rohm&Haas
Item 05: Tronox R-KB-2, Tronox Pigments GmbH
Item 06: Talc AT, Norwegian Talc A/S
Item 07: Milicarb, Omya GmbH
Item 08: Bentone 38, Elementis Specialties
Item 09: Desmodur L75, Bayer

Table 4.4: Guideline formulation of a conductive waterborne primer[3]

Item	Raw material **Component 1**	wt.-%
01	Polyester-polyurethanedispersion	32.0
02	Aerosil	1.7
03	Carbon black, conductive	1.3
04	Bentonepaste 3%	6.0
05	Butyl glycole	1.5
06	Surfactant solution 52%	1.5
07	Titanium dioxide	23.0
08	Barium sulfate	13.0
09	TalkumTalc	3.7
10	Emulsifier	1.1
11	WasserWater	10.2
12	Xylene	1.0
	Component 2	
13	Cyclo aliphatic polyisocyanate	5.7
	Total	101.7

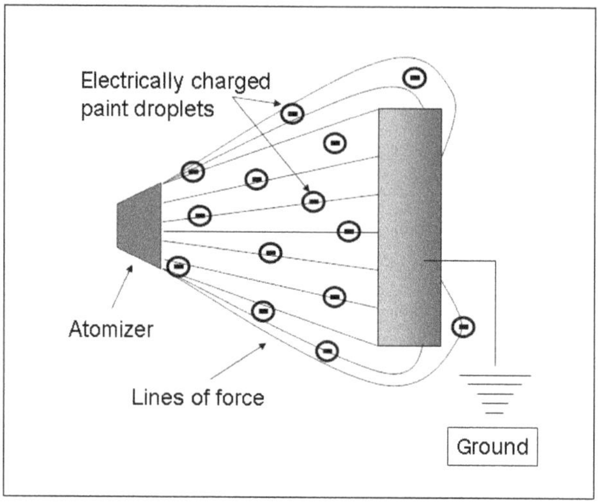

Figure 4.4: Schematics of electrostatic coating

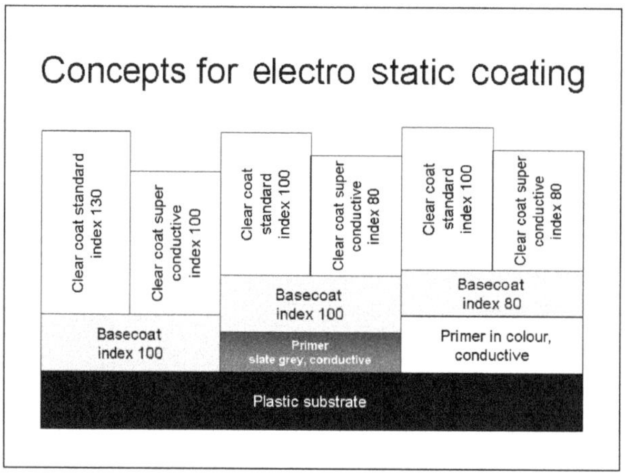

Figure 4.5: Various two and three layer ESTA coating concepts (index refers to paint consumption, 100 is the reference)

Two component primers are often cross-linked with HDI-type aliphatic isocyanates (hexamethylene diisocyanate), and in some cases aromatic isocyanates can also be used for solventborne 2K systems to improve curing speed and performance. As with IPDI based isocyanates, improved curing speed can be observed, however particular attention has to be directed to maintaining necessary flexibility.

The solvents must be chosen with regards to the substrate that shall be painted. A certain swelling of the substrate can actually promote adhesion of the primer. Even waterborne primers usually contain small amounts (<10 %) of co-solvents that can have adhesion promoting effects and are necessary for proper film formation. If the interaction between solvents and plastic matrix is too strong embrittlement and even cracking, caused by the release of internal stress from the moulding of the parts, can be the consequence.

Special additives for primers promote substrate wetting and adhesion. In addition all generally used coatings additives like levelling agents, dispersants, defoamers or catalysts are possible ingredients. When choosing the additives it's always necessary to study not only the desired effects but also the impact on adhesion and intercoat adhesion under process and service conditions.

Primers and fillers are normally pigmented with titanium dioxide, carbon black and functional extenders. The extenders are used to control mechanical properties (e.g. stone chip resistance), adhesion and intercoat adhesion and to support the barrier effect against solvents and inward water diffusion under humid climate conditions.

Primers are made conductive by the use of conductive pigments (carbon black, doped materials). Conductive primers allow the electrostatic coating of the non-conductive plastic. In the electrostatic coating process the coating is electrically charged at the atomizer (pneumatic or high speed rotation), the work piece is grounded. The charged paint droplets move along the electrical field lines to the part where the static is discharged to the ground (see Figure 4.4). This coating method dramatically improves the transfer efficiency by reducing the overspray. The conductive primer prepares the otherwise non-conductive plastic part for this process.

Nevertheless nowadays electrostatic coating on plastic without conductive primer is known. One utilizes the high self-conductivity of waterborne coatings or adjusts the wet film conductivity of solventborne clear coats by additives and solvents to low electrical resistance. The wet coating film itself can discharge the static in a “dynamic grounding”. Otherwise new arriving paint droplets would be repelled due to the same sign of paint and painted surface.

The different coating concepts result in varying material consumption. This is shown as indexed comparison in Figure 4.5. Standard is a three layer system with a conductive, slate grey primer. The basecoat in this example is always waterborne, that’s why the consumption is set to 100 with and without conductive primer. With the colour coded conductive primer the amount of basecoat can be reduced because of the colour consistency of primer and base. If an adhesion additive is mixed into the basecoat the addition must be compensated by higher film builds. The standard clear coat with wet film conductivity in a range of 300 to 500 kΩ is less transfer efficient in a two layer approach because the discharge through the clear coat layer is not good enough. Super conductive clear coats with electrical resistance <100 kΩ compensate for this drawback.

Even if primers are often substituted by primer-free coatings systems they still have their right to exist. On plastics on which it’s difficult to obtain adhesion, on solvent-sensitive substrates or under basecoats with limited hiding power a three layer system comprising primer, basecoat and clear coat can be the more economic and qualitatively better alternative.

4.4 Topcoats, mono-coat systems

Topcoats are the top layer of a two or multilayer system comprised of primer plus an additional coating layer to provide protection against the elements. Then there are mono-coat systems, topcoats that are applied directly on the plastic substrate, not requiring a primer layer. Both topcoat types focus the decorative and economic protection of the painted parts. The bandwidth of visual effects reaches from low gloss contrasting colours in all shades of grey that laymen usually do not perceive as painted parts, through single-layer pearlescent or metallic effect coatings to high gloss solid shade 2K topcoats. All of these paint systems are in use as solvent or waterborne topcoats or mono-coat systems in all areas of painted plastic parts – from television cabinets to automotive exterior or interior. In consequence the specifications and the coatings formulations are as manifold as their end use. Abrasion resistance (furniture, kitchen equipment, audio devices), resistance against food stuffs (kitchen, automotive interior), cosmetics ingredients (electric shavers, auto interior, bath room furniture), cleaning agents of all types, against hot fat (kitchen) or suds may be of interest as much as weathering resistance, stone chip resistance or elasticity.

1K solvent or waterborne coatings are often found as single-layer systems in the manufacture of consumer electronic components. Resistance to substances and cleaning agents of domestic environments in addition to the always expected excellent adhesion to different plastics are key design criteria. More exposed parts in outdoor use or for automotive applications are painted with tried and tested 2K solventborne coatings besides increasingly used 2K waterborne systems. In automotive interior market segments, the share of waterborne systems in Germany reaches more than 90 %, world wide it might be over 60 %. Table 4.5 represents typical characteristics of plastics topcoats.

Table 4.5: Typical characteristics of plastic topcoats

	Topcoats for plastics			
	Waterborne		Solventborne	
Colour/effect	contrast colour grey-black	effect colour pearlescent, metallic*	contrast colour grey-black	effect colour pearlescent, metallic*
Solids [wt.%]	25 to 40	20 to 30	30 to 45	25 to 35
Binder-to-pigment ratio	approx. 1 : 0.4	1 : 0.3 to 1 : 1	approx. 1 : 0.5	approx. 1 : 0.3
Density [g/cm³]	1.0 to 1.1	approx. 1.05	approx. 1.03	approx. 0.97
Flow time [s] DIN4 cup	19 to 45	25 to 50	18 to 22	20 to 35
Drying/curing	minimum 30 minutes 80 °C		minimum 30 minutes 80 °C	
Film build [µm]	20 to 30	15 to 25	20 to 35	20 to 30
Gloss [units @ 60°]*	2 to 20	5 to 15	2 to 35	10 to 55

*: with limitations

Table 4.6: Topcoat guideline formulations with varying flexibility[4]. The elasticity diminishes from A through C.

Item	Raw material **Component 1**	wt.-% A	wt.-% B	wt.-% C
01	Polyester, OH-functional	22.6	20.1	15.1
02	Polyacrylate, OH-functional	9.1	12.4	18.5
03	Leveling and slip additive, 10 % in butyl acetate	0.8	0.8	0.8
04	Catalyst 10 % in butyl acetate	0.8	0.8	0.8
05	HALS 10 % in 1-methoxypropylacetate-2	3.8	3.8	3.8
06	Titanium dioxide	22.8	22.8	22.8
07	Bentone 38; 10 % intermediate	3.8	3.8	3.8
08	1-methoxypropylacetate-2	6.8	6.5	6.1
09	Solvent naphtha 100	7.0	6.5	5.6
10	Diacetone alcohol	3.9	3.9	4.1
	Component 2			
11	Aliphatic polyisocyanate	18.6	18.6	18.6
	Total	100.0	100.0	100.0

Item 01: Desmophen 670, 80 % in butyl acetate, Bayer
Item 02: Desmophen A 365, 65 %, Bayer
Item 03: Baysilone paint additive OL17, OMG Borchers GmbH
Item 04: Dabco 33LV, AirProducts
Item 05: Tinuvin 292, Ciba Spezialitätenchemie
Item 06: Tronox R-KB-2, Tronox Pigments GmbH
Item 07: Elementis Specialties
Item 11: Desmodur N75, Bayer
The elasticity of the top coats decreases from A to C

4.4.1 *Resins for topcoats*

The most important difference in topcoats for plastics to those used on metal substrates is the choice of resins used in the design of the specific coatings. As shown in more detail in Chapter 4.6.1.1 the mechanical behaviour of the coating must be similar to the characteristics of the plastic being used. If this is not respected, coating may negatively impact component performance. As multifaceted as the potential applications the choice of resins in use becomes. 1K systems based on solventborne polyurethane elastomers with relatively low solid content, cellulose acetate butyrate (CAB) or high molecular weight acrylics stand besides 2K materials based on hydroxyl functional polyesters or polyacrylates, cross-linked with aliphatic isocyanates which are the norm for designing such chemistries. Waterborne coatings are dominated by 1K and 2K polyurethane dispersions. Almost all resin suppliers have guideline formulations for many application examples available.

Relatively new approaches to formulate topcoats are pigmented UV systems. The challenge in these coatings is the competitive situation between light absorbing pigments and photo initiator. For UV curing topcoats, hiding pigmentation is critical because it's necessary to use light scattering pigments to greater extent that prevent the UV light from advancing to the deeper paint areas. As a result cross-linking is not sufficient in the depth of the coating. Only a few microns can be well cured. The same effect has the absorption of UV light what means that even under favourable conditions film builds of not more than 20 µm are realistic. Actually efforts are under way to improve the through curable film build by modifying the initiator systems [5].

4.4.2 *Pigments and fillers*

For the wide spread contrast topcoats, standard carbon blacks in combination with rutile type titanium dioxides are used. For the formulation of high gloss coatings, one uses the whole range of inorganic and organic colour pigments. They are chosen dependent on the intended field of application and the associated light and weather fastness combined with optimum economical efficiency. In this regard there is no difference between plastic coatings and their equivalent used on other substrates. The formulation of waterborne or solvent-based pearlescent and metallic effect colours follows the same basic principles used in basecoats (see Chapter 4.5). Special attention must be paid to the protection of the effect pigments by the resin matrix.

Fillers are able to improve important characteristics such as adhesion, chemical or stone chip resistance. The fillers often used include talc (hydrated magnesium silicate), barium sulphate or chalk (calcium carbonate). Talc is said to have the characteristics, based on its platelet form, to distribute eventually acting impact forces in the layer (energy dissipation) and to prevent the film from damage by this mechanism.

4.4.3 *Additives for topcoats and mono-coat systems*

A frequently used family of additives are the so-called **matting agents**, finely dispersed solids that allow controlling the gloss level of the thereby modified coatings. By far the most widely used matting agents are based on pyrogenic or precipitated silica with particle sizes of just a few microns that can have organic or inorganic surface treatments for stabilization. Also known are organic polymer matting agents made from polyurethanes, polycondensates, polymethyl methacrylate (PMMA) or silicones. Usual fillers like talc can serve as matting agents within some limits as well. Primary objective of each type of matting is a uniform gloss level under all conditions of application and curing what is in most cases not perfectly achieved. The matting agents are subject to the same internal flow processes in the coating layer as the other ingredients. Only the particles direct in contact with the surface really have matting character. This makes it simple to understand that layer thickness as well as flash-off conditions directly influences the gloss level. Matting agents are of higher density than the surrounding resin/solvent matrix. Therefore they tend to follow gravity during flash off. Higher film builds or longer flash-off times as well as heavier solvents promote this effect and usually result in higher gloss levels. Surface treatment of the matting agents is, under others, aimed to reduce this effect.

Matting agents have large outer (by particle size) and inner (by porosity) volume so they bind large part of the resin what is in consequence only to a limited extent available for adhesion and stability. Therefore feasible gloss levels are limited in the low gloss range and should not be much less than 10 gloss units at 60° angle for high quality automotive exterior paints. Interior applications or consumer electronics sometimes allow lower levels. It has to be kept in mind that scratch and mar by fingernails or metal objects becomes more sensitive at lower gloss levels.

Further additives for topcoats are slip and levelling agents based on silicones and acrylates, light stabilizers (HALS, sometimes UV absorbers) and catalysts or wax additives to improve surface sensitivity. The influence of the additives on the overall performance, as in the case of levelling agents or waxes must be tested and confirmed regarding re-coatability and intercoat adhesion.

4.5 Basecoats

Basecoats are the colour and effect giving component in a multilayer coating and are over coated with a clear coat for protection (see Figure 4.2). Originally this multilayer structure was used for effect colours only. Solid shades were realized using topcoats without further protective layer. In the 80s and 90s of the 20th century this multilayer approach prevailed for solid colours as well. This is justified by the much better long term stability and fitness for use of the compound of colouring basecoat and protecting clear coat. Added to this are process advantages for the user: if it comes to inclusions single-layer topcoats usually cannot be reworked without visual interference. During polishing the coatings layer thickness is reduced and it could be that it doesn't cover the underlying layer any more. This can result in a visible polishing spot. Multilayer systems have the advantage that only the non pigmented clear coat layer is polished.

Basecoats for plastics may have distinct differences in composition compared to OEM basecoats for car body applications in automotive assembly lines. The request for flexible systems and the usually lower curing temperatures are reasons for these differences.

4.5.1 Solventborne basecoats

Today solventborne basecoats are still used in many regions. In the US the waterborne alternatives still did not prevail. Instead so-called medium and high solid basecoats with solid contents of up to 40 % by weight, sometimes even more, are used. Despite the European VOC regulation (VOC = **v**olatile **o**rganic **c**ompounds) even in Europe solventborne basecoats are widely used. This is sometimes still true for job coaters and smaller automotive suppliers in Germany. In general the process advantages of these systems (fast flash-off without intermediate cure) must be compensated by higher efforts for air abatement. Usual solids for the solventborne basecoats used in Europe range between 25 % by weight for brilliant metallics or black up to 40 % by weight for light coloured solid shades. The basecoats are application dependent sprayed in two passes with pure pneumatic atomization or by combination of high speed bell atomizers and final air spray. The latter operation has

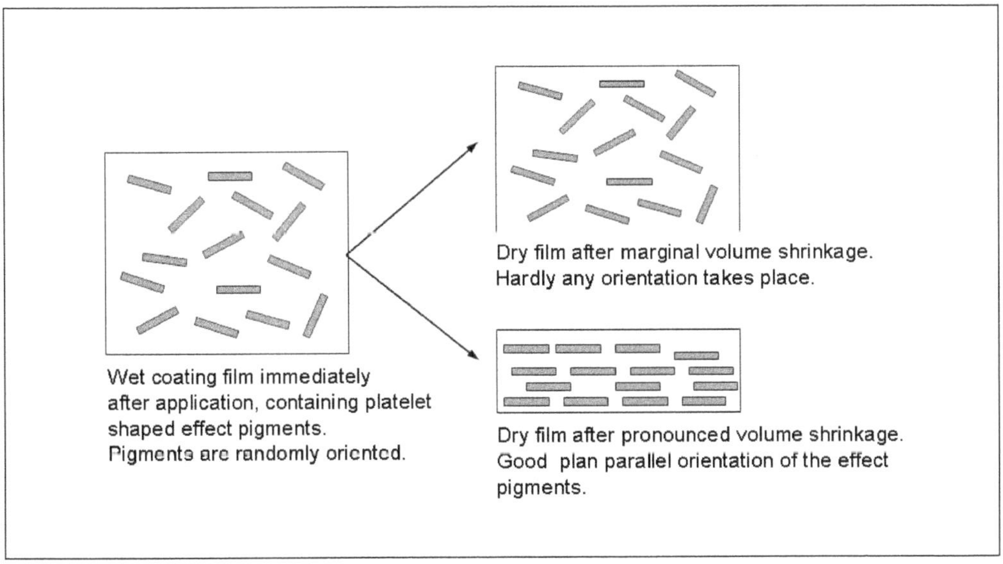

Figure 4.6: Effect pigment alignment by volume shrinkage

the advantage of a significantly reduced material consumption and a more homogeneous layer thickness distribution. Pre-condition is the application of conductive primers (see Chapter 4.3).

After a short flash-off the basecoat layer is over coated with clear coat. Both layers are baked together. The baking conditions range in most cases from 30 to 60 minutes at 80 °C. Coatings on temperature sensitive substrates are cured at lower temperatures, e.g. 60 °C.

Besides the effect and colour giving pigments the binders play the dominant role in the formulation of solventborne basecoats. Common formulas are based on cellulose acetate butyrate (CAB) in combination with polyesters, acrylic and melamine resins. The CAB part is responsible for the effect and prevents from re-dissolving by the clear coat's solvents. CAB is used in different viscosities. The higher the viscosity, respectively the molecular weight, the

Table 4.7: Guideline formulation solventborne basecoat[6]

Item	Raw material	wt.-%
01	Aluminum pigment 65% in solvent	4.0
02	Butyl acetate	6.0
03	Surfactant	0.3
04	EVA wax dispersion	5.0
05	Polyester resin	1.0
06	Acrylic resin	21.6
07	Defoamer	0.3
08	Methoxy propyl acetate	25.0
09	Butyl acetate	20.0
10	Ethyl acetate	16.8
	Total	100.0

Item 01: Stapa Metallux, Eckart GmbH
Item 03: Zetasperse 2300, Air Products
Item 04: Ceratix 8461, Byk GmbH
Item 05: Haftharz LTW, Evonik Degussa
Item 06: Degalan M890, Evonik Degussa
Item 07: Byk 392, Byk GmbH

Preparation:
Dispers item 01, 02 u. 03 (5 min.)
Add item 04 and dispers (5 min.)
Add item 05, dispers
Add item 06, dispers
Add item 07, homogenize
Add item 08 to 10, homogenize

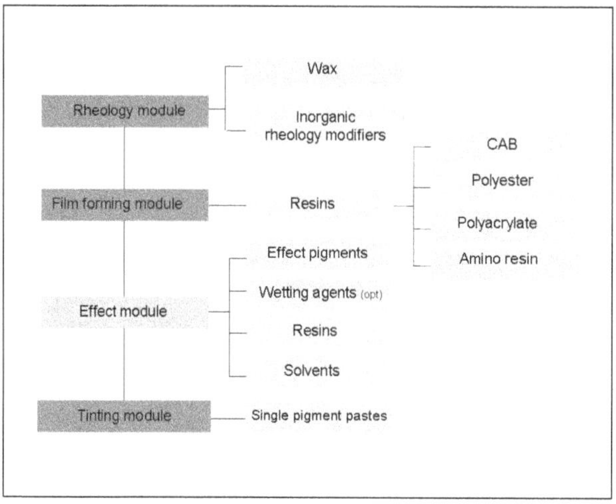

Figure 4.7: Schematics of solventborne basecoat composition

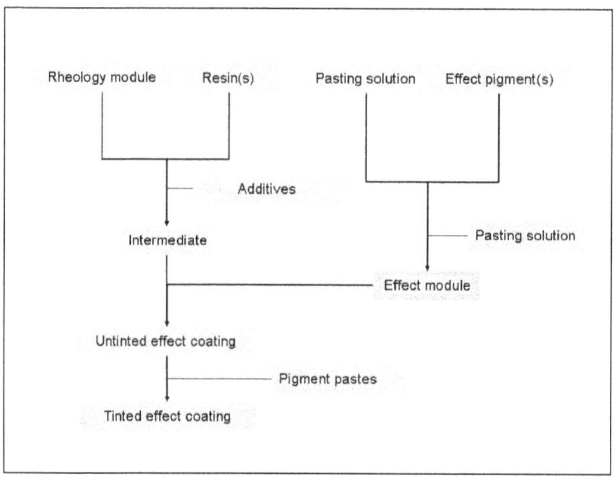

Figure 4.8: Schematics of solventborne basecoat production

lower is the resulting solid content and also the susceptibility to resolution by the clear coat. The limited resolubility might constitute a potential drawback: the clear coat adhesion can suffer on basecoats with too much high molecular weight CAB.

Low solids are important for the formation of optimum colour travel (flop) for brilliant metallic colours. A detailed discussion of flop characteristics and associated parameters can be found in the chapter dealing with effect pigments for waterborne basecoats (Chapter 4.5.2.3). As shown in Figure 4.6 significant volume shrinkage caused by the low solids occurs during drying. This forces the metallic particles in a position of parallel alignment.

This is supported by proper adjustment of the evaporation rate of the solvents in the basecoat formulation. The flash-off time must not be too short to allow for pigment orientation but in the same time not too long: already aligned particles can be displaced from their optimum position.

The co-binders often used control adhesion, levelling and in automotive formulations also stone chip and humidity resistance. The weather fastness has to be observed when it comes to basecoats for exterior use.

Normally basecoats are not produced in one process from ground up but are made from intermediates. According to their function, they can be classified in modules: film forming, rheology, effect and tinting modules. The composition of a solventborne basecoat based on such modules is shown in Figure 4.7. The schematics of the production process are illustrated in Figure 4.8. First of all the effect pigments have to be incorporated by means of wetting the single pigment particles with solvents so that they are available in form of mono disperse particles [7]. This is essential to achieve full efficiency. The additionally used colour pigments are most times added as pigment pastes. This assures that the pigments have constant quality in regards of colour position, colour strength and transparency. Colour adjustment is increasingly supported by multi angle colorimeters. Visual inspection as final control step is often still state of the art.

4.5.2 *Waterborne basecoats*

Since the beginning of the 1990s, starting in automotive body coating, waterborne basecoats are increasingly used. With a certain delay the automotive suppliers for plastic parts in Europe and especially in Germany started to use waterborne basecoats in their paint lines.

The application of waterborne basecoats is in some points significantly different to the solventborne type. All assembly units of the paint line that come into contact with the liquid paint like storage tanks or pipes must be made of stainless steel to prevent corrosion of the installation and contamination of the basecoat. The electric conductivity of waterborne basecoats needs to be considered when electrostatic application shall be used. Several options are available:

- External charging with the disadvantage of lower transfer efficiency and larger atomizers
- By so-called insulated design of the paint supply with the disadvantage that the paint supply is under voltage during operation. Costly safety measures are necessary.
- By using so-called potential separators between paint supply and application unit with the disadvantage of complex paint supply systems that have been relatively susceptible to malfunctions until recently.

The spray booths must be air conditioned to obtain reproducible paint results with regards to colour, effect and levelling. Standard conditions are 23 ±1 °C and 65 ±5 % relative humidity. Waterborne basecoat films must be conditioned to <10 % residual humidity prior to clear coat application. This is reached by curing or forced flash-off at a minimum of 10 minutes at 80 °C. If this is not respected the subsequent clear coat can show reduced gloss or haziness. Clear coat drying can be significantly delayed; the chemical resistance of the system does not reach its optimum and it could result in increased sensitivity against humidity, temperature or weathering.

Table 4.8: Waterborne basecoat for plastics

Item	Raw material	wt.-%
	Component 1	
01	Acrylic dispersion	251.3
02	Acrylic dispersion	26.1
03	DMEA, 10 % in demineralized water	6.5
04	Polyurethane dispersion	75.7
05	Demineralized water	42.2
	Stirr slowly for 5 minutes and add:	
	Premix of solvent	
06	Butylglycol/demineralized water 1:1	102.6
	Component 2	
07	Aluminum paste	107.5
08	Polyester polyol emulsion	85.1
	Viscosity adjustment	
09	Demineralized water	66.9
	Aluminum paste	
10	Aluminum pigment 65 % in solvent	36.7
11	Polyester resin	32.4
12	Butyl glycol	30.4
13	DMEA pur	0.5

Item 01: Setaqua 6801, Nuplex Resins bv

Item 02: Setaqua 6802, Nuplex Resins bv

Item 04: Bayhydrol VPLS 2952, Bayer

Item 08: Setaqua X 11583, Nuplex Resins bv

Item 10: Stapa Hydrolac WHH 8154, Eckart GmbH

Item 11: Setal 6306 SS-60, Nuplex Resins bv

Preparation

Mix component 1 and 2, stabilize for at least 16 hours. Stirr slowly for 10 minutes and adjust pH to 7.4 with DMEA (10 % in demineralized water). Adjust to spray viscosity 32 s DIN-4 with demineralized water.

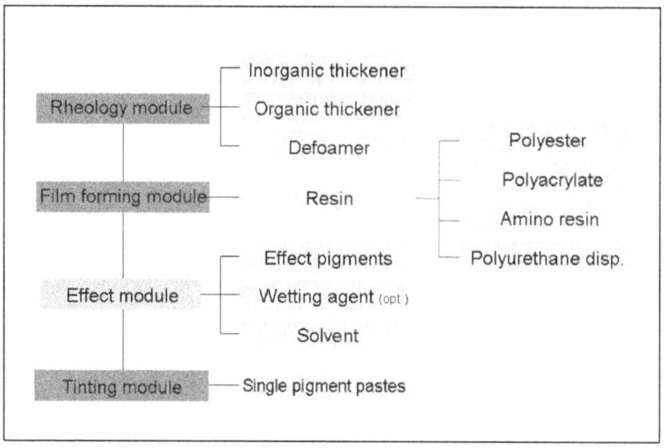

Figure 4.9: Schematics of waterborne basecoat composition

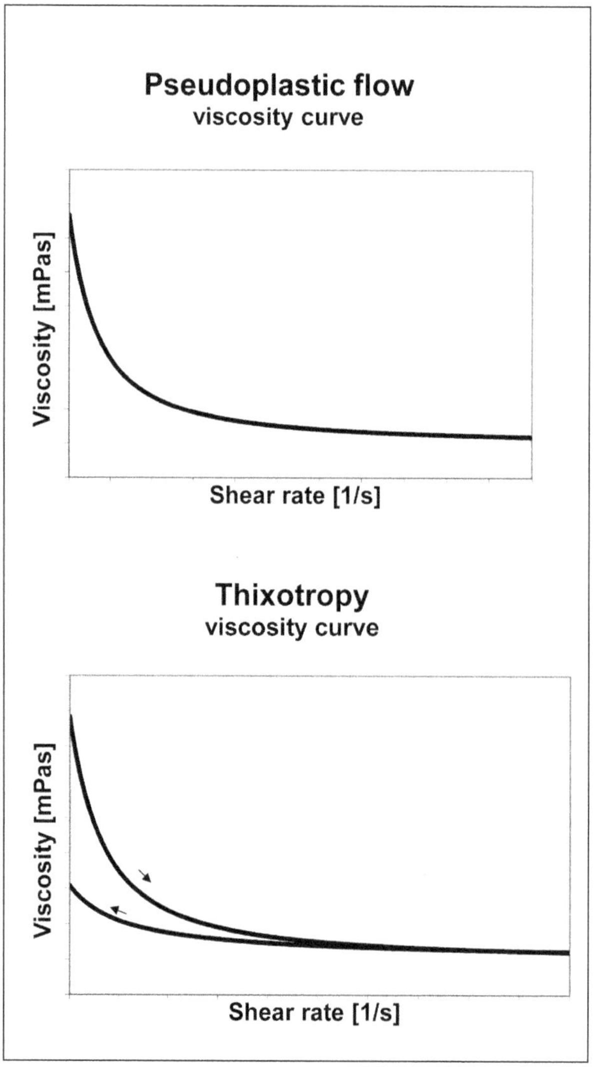

Figure 4.10: Pseudoplastic and thixotropic flow behaviour

The formulation of waterborne basecoats is in general more complex than the formulation of solventbornes. Waterborne basecoats for plastics are, due to the limited curing temperatures not higher than 80 °C for the entire coating system, even more challenging for the coatings developer. Today's state of technology in the automotive industry is one single formula for car body painting in the manufacturer's assembly line and for painting plastic parts at the supplier. A guideline formulation for this specialized type of basecoats is shown in Table 4.8. A diagram of the raw materials used in waterborne basecoats and their combination in modules is displayed in Figure 4.9.

4.5.2.1 *Resins for wateborne basecoats*

In the simplest case the film forming module can be polyurethane and/or polyacrylic dispersion only. The advantage of such formulations is excellent adhesion control even in two layer systems without primer layer, very good performance, e.g. good stone chip resistance for automotive applications, and a fast physical drying that prevents the basecoat from strike in by the clear coat's solvents. The disadvantage of this formulation design can be the fast, non reversible drying of the dispersion. In high speed bell applications paint can dry on the surface of the bells. Dry paint particles can then be expelled creating surface defects caused by paint inclusions. Even during production, storage and paint supply precautions against unwanted drying and self contamination are necessary.

Another formulation design principle is the combination of water-soluble, reactive polyesters with waterborne melamine resins and polyacrylate and polyurethane dispersions. Such formulas are also used for the coating of automotive bodies. Special attention has to be paid to the necessary flexibility of such versatile basecoats used for car body and plastic parts. Another challenge is the cross-linking reaction of polyester and melamine resin under the limited curing temperatures in plastic applications. If this doesn't work poor humidity resistance is just one of the adverse effects.

4.5.2.2 *Rheology control of waterborne basecoats*

Rheology of waterborne basecoats plays an important role during storage and especially during application of the material. Without rheological additives the coatings usually exhibit very low viscosity so the pigments would tend to strong sedimentation. Platelet-shaped effect pigments form a deposit that hardly can be re-dissolved. This often results in pigment agglomerates that make even high-grade effect pigments look grey and unsightly.

Such coatings would have a distinct sagging tendency. This can lead to bad effect formation by running and micro sagging, a film defect that appears just before visible sagging starts. Waterborne basecoats must be formulated with appropriate rheology modifiers that give them pseudoplastic (sheer rate dependent viscosity) or thixotropic (sheer time and sheer rate dependent viscosity) flow behaviour. Figure 4.10 illustrates these rheological fundamentals.

Waterborne basecoats are not much different in solid content compared to their solventborne analogues. The statement of Chapter 4.5.1, Figure 4.6, about pigment orientation by volume shrinkage should thus be valid – if there would not be water as dominating solvent with its special characteristics (see Table 4.9). In particular the slow evaporation rate of water initially averts this mechanism. The rheological adjustment of the basecoat supports an optimum effect formation by securing a homogeneous effect pigment distribution in the paint film until the volume shrinkage finally takes place.

Table 4.9: Properties of water as solvent

Solvent	**Water**	**n-Butanol**	**Butyl glycole**	**Xylene**	**Methoxy-propanol**	**Dipropylene glycole dimethylether**
Boiling range [°C]	100	117	167 to 173	136 to 139	120	175
Evaporation rate (diethyl ether = 1)	80	33	160	14	25	95
Heat of vaporization [kJ/kg]	2258	599	368	344	431	257
Surface tension [mN/m]	72	25.5	27.8	29.5	28.3	26.3
Density [g/cm³]	1	0.81	0.9	0.87	0.92	0.9
Molar mass	18	74	118	106	90	162

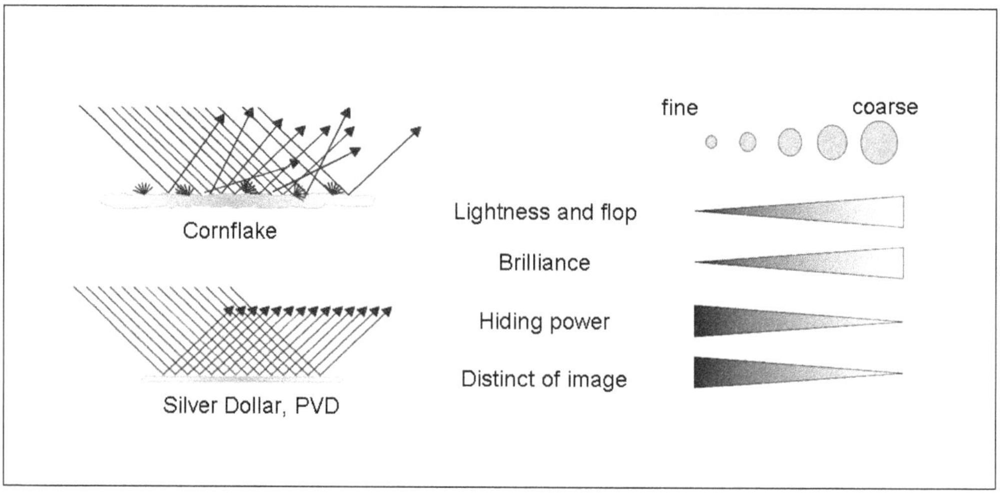

Figure 4.11: Optical characteristics of metallic effect pigments *Source: Eckart GmbH, Velden, Germany*

4.5.2.3 *Effect pigments for waterborne basecoats*

According to DIN 55943 effect pigments award to the medium applied additional characteristics besides colour like e.g. angle dependency of colour and gloss. The most important effect pigments for coatings are metallic and pearlescent pigments. Metallic pigments are lustre pigments that align parallel to the surface and generate metal like gloss by reflection of light. Their predominant representatives are platelet-shaped aluminium pigments (aluminium flakes). Pearlescent pigments can provide colour by optical interference.

The optical properties of metal effect pigments can be described as a combination of directed reflection of the incoming light on the metallic pigment's surface and light scattering on the edges and uneven areas of the surface of the pigment (see Figure 4.11) [7].

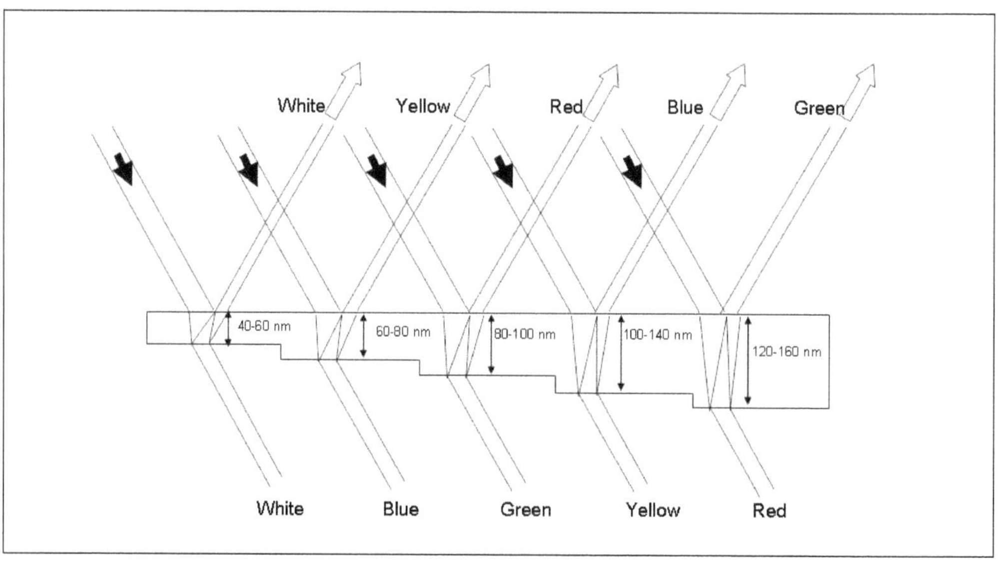

Figure 4.12: Interference colour as a function of the optical path length [10]

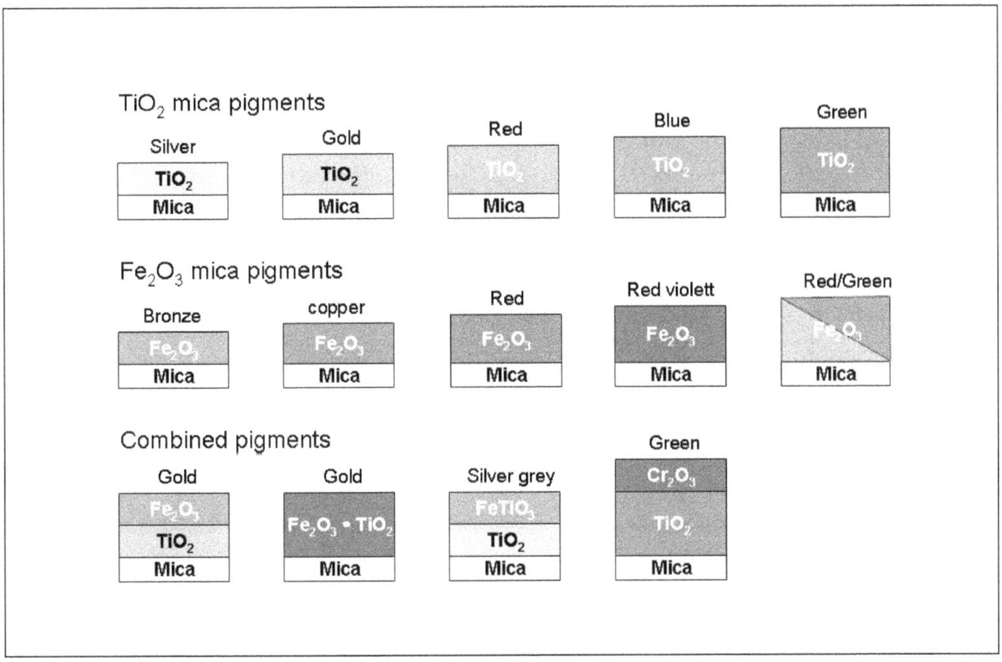

Figure 4.13: Effect colour as a function of metal oxide thickness and combination [9]

In addition particle size and shape determine lightness, flop or travel characteristic, brilliance and hiding power of the chosen effect pigment. The coarser the pigment and the rounder the platelet's shape and structure are, the percentage of reflection and thereby the metallic brilliance and lightness increases. Likewise the flop effect is more pronounced, that means the lightness travel as a function of the observing angle [7].

Especially aluminium pigments cannot be easily used in waterborne (base)coats. Aluminium reacts with water to form aluminium hydroxide and hydrogen. Due to the large surface area of aluminium pigments this reaction can be very intensive. Even at slow reaction rates loss of gloss and shelf life impairments (agglomeration) are the consequences [7]. Different methods to protect the aluminium pigments are used and must be selected depending on the paint system design. The most dominant procedures are the use of corrosion inhibitors, forming of conversion layers on the pigment's surface or the complete encapsulation of the aluminium pigment. For the system dependent selection not only the corrosion stability in the formula but also application criteria like optical characteristics or humidity resistance of the final coating are of importance.

Pearlescent pigments comprised of transparent platelets with a high refractive index are made by coating the natural or synthetic mica with metal oxides. When oriented plane-parallel, they exhibit characteristic pearl lustre by multiple reflection. Pearlescent pigments that show additional interference colours are called interference pigments [9]. The thickness of the metal oxide layer is decisive for the resulting colour impression (Figure 4.12).

As shown in Figure 4.13 the selection of metal oxides and combinations of subsequent layers make additional effects available.

Pearlescent pigments are used in many decorative applications and are widely applied in plastics coatings for consumer electronic devices, high quality cosmetic packaging or for automotive parts.

4.6 Clear coats for plastics

Clear coats are transparent and usually colourless coating materials. As the final layer of the coating system, they protect the underlying layers from the elements of weather, chemical attack, influences of processing liquids and mechanical damage. In special applications clear coats are applied for protection directly to the plastic substrate.

Table 4.10 illustrates the possible usages of clear coats in the painting of plastics:

- Final layer in a three layer system consisting of primer, basecoat and clear coat
- In a primer-free system with basecoat and clear coat only
- As clear coat directly on mass coloured or otherwise decorative substrate, eventually combined with transparent adhesion promoter

Different technologies are used to formulate clear coats:

- 1K or 2K clear coats
- Solventborne
- Waterborne
- Dual cure UV clear coats
- Mono cure UV clear coats
- Powder clear coats
- Sol/Gel clear coats
- Nano modified clear coats

Table 4.10: Application of clear coats based on technologies

Technology	Typical curing temperature	Actual applications	Possible future applications
High bake 1K solvent-borne clear coats	120 °C	Automotive parts, USA	
2K solventborne clear coats	80 °C, 60 °C	Automotive parts, European OEMs Consumer Electronics General industry	
1K clear coats	e.g. room temp.	Low budget applications	
2K waterborne clear coats	80 °C	Automotive parts, mainly EU	
Mono cure UV	UV	Cell phones, CD, ski, snowboards, ...	Automotive parts
Dual cure UV	80 °C + UV	Automotive parts	Automotive parts General industry
2K solventborne nano clear coats"	80 °C		Automotive parts General industry
Powder clear coats	140 °C	Temperature resistant automotive parts	Temperature resistant automotive parts
Sol/gel clear coats	120 °C	Headlamp lenses Polymer glazing	

Clear coats can be adjusted to various gloss levels from high gloss to matt finishes, even soft feel clear coats are known.

In most cases the flexibility of the clear coat formulation needs to be accustomed not to reduce the impact resistance of the substrate too much. Admittedly all layers contribute to the total system performance, but the clear coat has a significant part in the reduction of the impact behaviour. High flexibility on the one hand has to meet with excellent resistance as well as sand ability and polish ability on the other hand. Actual requirements to further improve scratch resistance confront the developing chemist with extra challenges.

4.6.1 Property profile of clear coats

Next to the standard characteristics like solid content or viscosity clear coats face, is the importance to wet out surfaces at low film builds and to have high sagging limits. By optical reasons (brilliance, depth) and to improve the overall resistance clear coats are applied in a range of 25 to 45 microns. The obliged application window is even wider.

The **wetting limit** is - at standardized substrate and application - the film build that just shows an even surface without discrete droplets.

Sagging limit on vertical surfaces is determined by using special panels or critical original parts. Substrate and application must be standardized and defined. Another approach to understand this characteristic is the measurement of the viscoelastic behaviour of the clear coat over time before, during and after sheer stress. Methods for clear coats in application viscosity are still under development. Rheometers with high resolutions over a large sheer rate area are necessary for such studies.

It is self evident that the aim is the lowest possible wetting limit combined with the highest feasible sagging limit; two targets that interact with each other. The rheological characteristics of a clear coat are, besides other influences, decisive regarding the possible levelling (appearance) of a clear coat film. Levelling and surface structure are subject to measurement [11].

4.6.1.1 Hardness, scratch resistance, sanding and polishing

The clear coat itself can be from flexible to hard. This has consequences for the total system of plastic and coating. The flexibility of the plastic-coating-compound can be measured

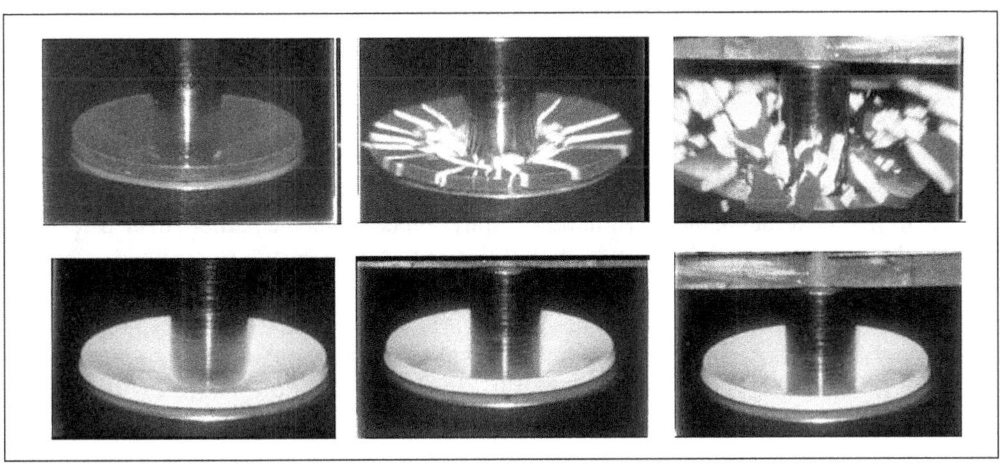

Figure 4.14: Ductile (below) and brittle (above) failure mode Source: Bayer Materialscience, Leverkusen, Germany

dynamically by impact tests (e.g. Charpy, DIN EN ISO 179, IZOD, DIN EN ISO 180, instrumented impact test ISO 6603-2 or front and rear impact ECE R 42) or more static by cylindrical or conical mandrel bending test. This is usually done starting at room temperature progressing to -30 °C in steps of 10 K. The ductile to brittle transition temperature is the aim of the investigation respectively the first occurrence of cracking (see Figure 4.14). Focus is as low as possible values which can't be lower than those of the substrate used.

The inherent **surface hardness** of clear coats can often be misleading. It's obvious that harder clear coats usually have better **chemical resistance**, and scratch resistance is a direct function of surface hardness. What is surprising: flexible clear coats also show excellent **scratch resistance** due to their reflow or self healing properties. Clear coats in a medium hardness range are susceptible to scratch and often don't have reflow properties. Surfaces, hard like glass, are almost not scratchable. Hard, inorganic nano particles are incorporated intended to build such surfaces. As conclusion scratches can be prevented by very hard clear coats ("diamond"), highly elastic ones ("rubber") or by clear coats with low glass transition temperature (T_g) exhibiting good reflow characteristics.

Balancing hardness, polish ability and chemical resistance

Each single strategy shows drawbacks. Very hard clear coats make the full system brittle. Highly elastic formulations need extra efforts when it comes to sanding and polishing. Clear coats with low T_g are not resistant enough against chemical attack. By this reason the formulation strategies must be combined in a way to create the best possible balance of properties.

In the painting process sanding and polishing abilities of clear coats are of high importance. Inclusions in the clear coat are levelled by sanding. The resulting sanded spot becomes invisible through polishing. The expenditures for sanding and polishing shall be as minimal as possible by economic reasons. Sanding shall happen fast and on the smallest feasible spot. The clear coat is inevitably heated up and the glass transition temperature could be eventually exceeded. Clear coat sticks to the sanding paper, heavy sanding marks can be the result. Subsequent polishing is usually done at room temperature with suspensions of very fine sanding media. By warming the clear coat layer the polishing agent can penetrate the soft while warm clear coat and result in unwanted polishing spots, so-called holograms. The glass transition temperature of the coating film combined with its surface hardness controls this behaviour. Clear coats in a mid hardness range show advantages in this topic. Highly flexible clear coats are polished using liquid nitrogen to constantly cool the surface to temperatures below T_g.

4.6.1.2 Chemical resistance

Clear coats as interface to the environment are exposed to various chemical attacks. Daily life household effects must withstand aggressive cleansing agents, notebooks or mobile phones are in constant contact with hand perspiration or cosmetics, other surfaces can be attacked by beverages and foodstuffs. In addition automotive parts are strained by working materials like gasoline, diesel, RME (rape seed oil methyl ester, biodiesel) or brake fluid and environmental impacts like acid rain, bird droppings or tree resin. Whether and how strongly the clear coat is affected by these substances is, under others, a question of temperature. To determine the temperature threshold an indirect test method with painted metal panels and a temperature-gradient oven is widely used. At a given exposure time it is observed at what temperature first damages occur. Highly flexible plastic coatings usually have some degrees lower thresholds than harder alternatives.

4.6.1.3 Weather ability

The term weathering summarizes a bundle of potential deterioration mechanisms:

- Light (degradation by (UV) radiation)
- Humidity (hydrolysis, swelling)
- Acid rain (etching)
- Heat (thermo oxidation)
- Temperature cycle (stress induced by differences within the thermal expansion coefficients of substrate and coating or of the coating layers)
- Microbiological load (settlement of, damage by microorganisms)

Most clear coats are not affected by all deterioration mechanisms to the same extent. Clear coats on vehicles of all types or on window profiles must be equipped for the whole range. Clear coats for vehicle interiors or for plastic housings of mobile phones are mainly exposed to humidity, heat and temperature cycle. Even objects within the house can become unsightly by direct solar irradiation. Different resin technologies are damaged to different extents. Resins with a high content of aromatic structures will yellow faster than others with mostly aliphatic building blocks. Epoxy resins lose gloss very rapidly, finally chalking. Acrylic melamine clear coats are more prone to cracking than those based on 2K polyurethane chemistry. The potential protective mechanisms and the mode of action of the additives and resin technologies concerned are well studied for many years.

An important protective element is the absorption of UV light. According to the law of Lambert-Beer the absorption is related to absorber concentration and length of the light path or layer depth. Therefore UV absorbers primarily protect clear coats in deeper layers and the layers underneath but not the outer limits of the clear coat. Figure 4.15 shows the most important types of UV absorbers used for coatings.

UV absorbers rapidly dissipate the absorbed, destructive UV light as heat energy. The absorbed energy is transferred into vibration or rotation energy of molecular constituents. Requirement for the effectiveness of UV absorber is that this process is faster than a corresponding reaction in the substrate and that nor the UV absorber neither the protected polymer suffers damage during energy conversion [12].

2-(2-Hydroxyphenyl)-benztriazole	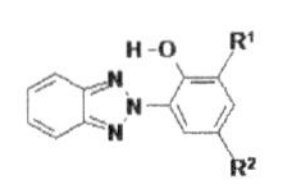
2-Hydroxybenzophenone	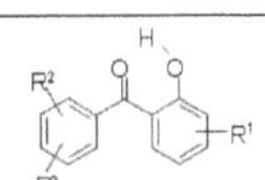
Hydroxyphenyl-s-triazine	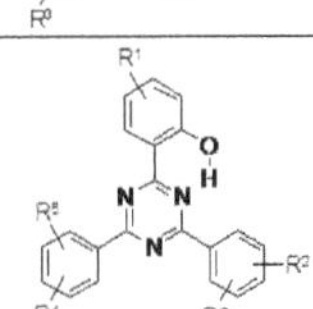
Oxalanilide	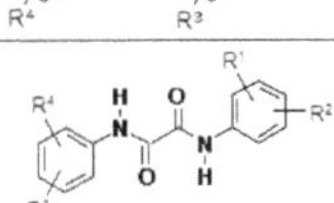

Figure 4.15: Technically important UV absorbers for coatings
Source: BASF Switzerland AG, Basel, Switzerland

The second component besides UV absorbers in a light stabilizing package is a **h**indered **a**mine **l**ight **s**tabilizer HALS based on 2,2,6,6-tetramethylpiperidine. The mechanism is based – simply put – on the effect that free radicals released by UV light are recombined to stable organic compounds and the photochemical degradation is interrupted. The

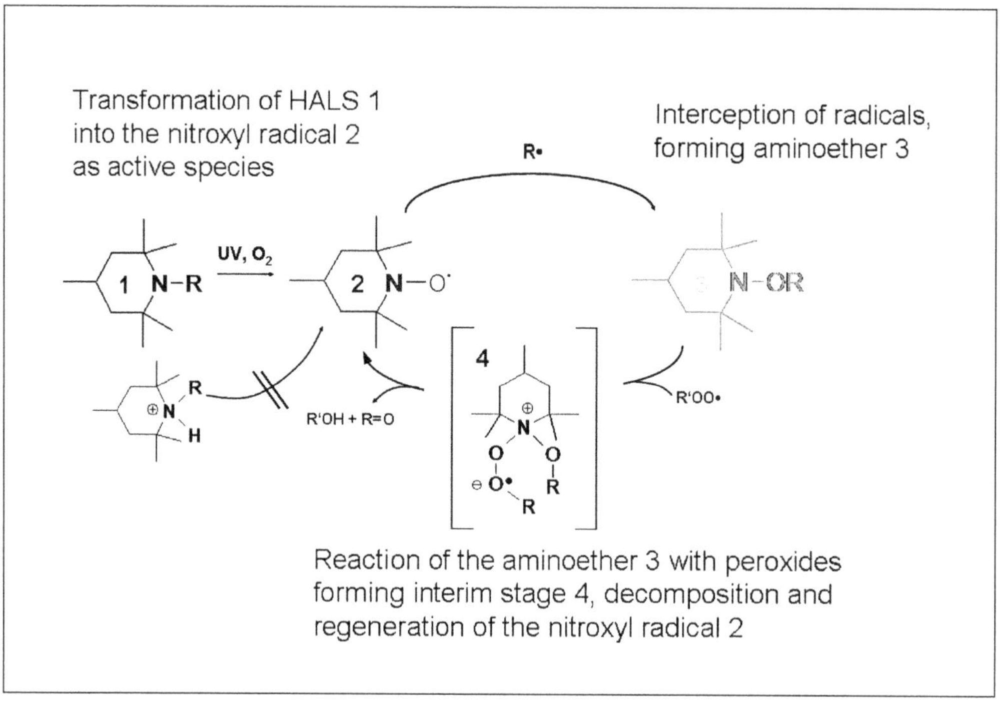

Figure 4.16: Denisov Cycle *Source: BASF Switzerland AG, Basel, Switzerland*

radical scavenger itself is re-established in a cyclic process. This process, known as Denisov Cycle is shown in Figure 4.16.

HALS can also improve instability against hydrolysis and oxidation. In some cases the use of additional antioxidants can be considered.

4.6.2 Formulating clear coats

4.6.2.1 Solventborne clear coats

The vast majority of clear coats actually used for the painting of plastic parts are systems based on organic solvents. Major reason is the excellent balance of economics, range of properties and application window. Solventborne clear coats for plastics are predominantly isocyanate cross-linked 2K polyurethanes. For the formulation of the base a huge variety of hydroxyl functional polyesters and polyacrylics is available. As a rule a combination of both chemistries is used. The more soft polyester serves for the flexibility of the system, the more rigid acrylic stands for stability and polishing properties. Of course also flexible acrylics and hard polyesters can be found in clear coat formulations, however these are less widespread. When choosing the resins the suitability for the intended use must be observed as mentioned above. Coatings for durable exterior use should have as little as possible aromatic containing structures.

As cross-linkers mainly pre-polymers based on hexamethylene diisocyanate (biuret or isocyanurate) come into question. Isophorone diisocyanates result in harder, more chemical resistant films. Both isocyanates are suitable for exterior use. The cross-linking density has to be chosen in a way that the desired properties are achieved reliably. NCO/OH ratios from 0.8 (less than equivalent) up to 1.3 (more than equivalent) are common practice. Excess isocyanate reacts with ambient humidity to polyurea, giving the film higher hardness and better chemical resistance.

To improve the sagging stability of clear coats rheology modifiers are used. By incorporating pyrogenic silica or crystalline urea [13] the flow characteristics become shear rate dependent (structurally viscous - pseudo plastic) and/or time dependent (thixotropic) (see Chapter 4.5.2.2, Figure 4.10). The rheology additives form internal structures that are temporarily destroyed by shear forces, the viscosity decreases considerably. This effect takes place during atomization of the clear coat and improves substrate wetting and levelling. After shearing the internal structure recovers more or less rapidly what prevents the coating from sagging on vertical surfaces.

Further additives are, dependent on the field of use, catalysts to accelerate the NCO/OH reaction, levelling agents based on silicones or acrylics, defoamers and - indispensable for exterior use - light stabilizers as described in detail in Chapter 4.6.1.3.

The applied solvents are partially determined by the chosen resins and isocyanates. Further solvents must be selected to allow for the best compromise between levelling, sagging resistance and substrate dissolution. In principle all coatings solvents are at the formulator's disposal. Table 4.11 shows typical characteristics of solventborne clear coats. Table 4.12 represents clear coat formulations of differing flexibility.

4.6.2.2 Waterborne clear coats

The European VOC regulation (VOC=**v**olatile **o**rganic **c**ompound) revived the development of waterborne clear coats for plastics in the last couple of years. Primers and basecoats were converted to waterborne systems many years ago. The remaining step that allows the further operation of older paint lines without expensive upgrades is the implementation of VOC compliant waterborne clear coats.

Hydro clear coats for challenging exterior applications are, as the solventborne standards, isocyanate cross-linked 2K systems. The basic resin composition follows the same principles as discussed for the solventborne formulations. The polymers have to be converted to the medium water appropriately. This can be realized by water soluble resins, externally or internally emulsified resins or through resin dispersions. The utilized isocyanates are fully or partially hydrophilically modified and usually premixed in water miscible solvents. Attention should be paid to the hydrophilicity brought into the system by resin and

Table 4.11: Key figures of clear coats for plastics

	Clear coats for plastics	
	Waterborne	Solventborne
Colour	colorlesss	colorless
Solids [wt.-%]	approx. 38 % @ application	45 to 50 % @ application
Binder-to-pigment ratio	n.a.	n.a.
Density [g/cm³]	1.02 @ application	1.00 @ application
Flow time [s] DIN4 / ISO4 cup	20 +/-3 DIN4/20 °C	28 to 35 s ISO4/20 °C
VOC	20%	50%
Drying/curing	≥25 to 30 minutes 80 to 85 °C object	≥25 to 30 minutes 80 to 85 °C object
Film build [µm]	25 to 45 µm	25 to 45 µm

Table 4.12: Guideline formulations for solventborne clear coats with varying flexibility[14]. The elasticity diminishes from A through C.

Item	Raw material **Component 1**	wt.-% A	wt.-% B	wt.-% C
01	Polyester, OH-functional	29.1	25.9	19.4
02	Polyacrilate, OH-functional	11.9	15.9	23.8
03	Leveling and slip agent, 10 % in butyl acetate	0.5	0.5	0.5
04	Accelerator, 1 % in butyl acetate	1.0	1.0	1.0
05	HALS, 10 % in methoxy propyl acetate	4.9	4.9	4.9
06	UV-Absorber, 10 % in methoxy propyl acetate	2.5	2.5	2.5
07	Methoxy propyl acetate	12.0	11.7	11.1
08	Solvent naphtha 100	12.0	11.7	11.1
09	Diacetone alcohol	6.0	5.8	5.5
	Total	79.9	79.9	79.8
	Component 2			
10	Aliphatic polyisocyanate	20.1	20.1	20.2
	Total	100.0	100.0	100.0

Item 01: Desmophen 670, 80 % in butyl acetate, Bayer
Item 02: Desmophen A 365, 65 %, Bayer
Item 03: Baysilone paint additive OL17, OMG Borchers GmbH
Item 04: Dibutyltin dilaurate
Item 05: Tinuvin 292, BASF Switzerland AG
Item 06: Tinuvin 1130, BASF Switzerland AG
Item 10: Desmodur N 3390, Bayer
The elasticity of the clear coats decreases from A through C

isocyanate, either by higher acid values of the binder, the emulsifiers or emulsifying concepts applied or the anionic stabilization of the system. The NCO/OH ratio is in any case significantly above 1 to compensate for the condensation reaction of isocyanate and water. This competitive chemical reaction does not have disadvantages for the film forming process. The network segments resulting from the condensation of isocyanate and water are durable and show good hardness and flexibility[13].

Additives used with waterborne clear coats are also light stabilizers, defoamers, levelling agents and catalysts. Attention should be paid to adequate compatibility of the additives with the medium water and the waterborne resins, not to neglect the hydrolysis stability[15].

The biggest challenge, next to the requirements of durability, is an application window reaching the one of the solventborne clear coats with the same levelling, anti sagging and high brilliance.

Up to date no systematic formulation concept for waterborne clear coats exists.

4.6.2.3 UV clear coats

The wavelength range of UV light is lying below visible light. The higher energy associated with the shorter wavelength allows using UV light for the curing of suitable coatings. Primary applications for UV curing paints on plastics are industrially coated floor coverings (made of PVC, rubber, polyolefin or linoleum), some automotive parts like polycarbonate (PC)

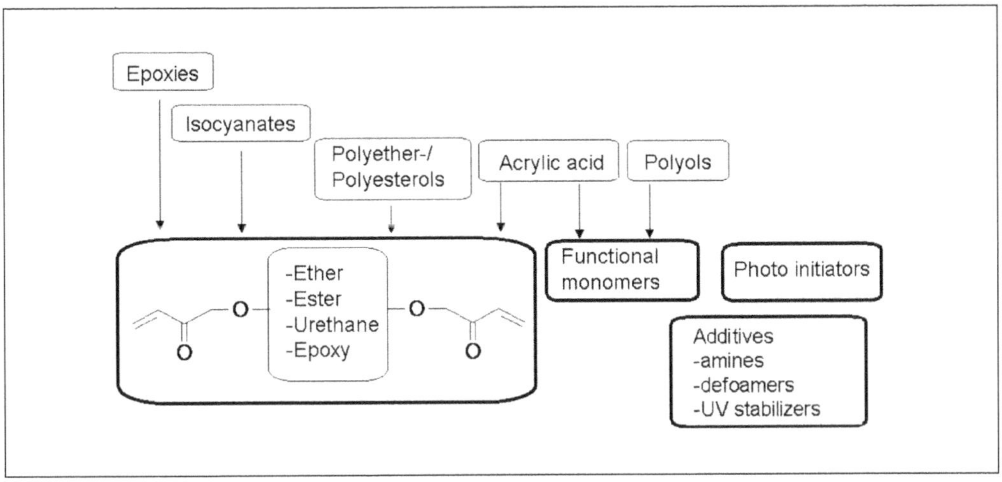

Figure 4.17: UV raw materials at a glance *Source: BASF SE, Ludwigshafen, Germany*

headlamp lenses, mobile phone covers made from PC or PC blends, plastic house ware as well as data storage devices such as compact discs (CD) or digital versatile discs (DVD) [16].

Striving for shorter processes and improved durability lead to the development of radiation curable paint systems for plastic parts. Radiation curable paints, generic term for paints cured by UV radiation or electron beams, are known since the 60s of the last century. Their only applications were flat substrates for the interior (parquet, furniture surfaces, panels) or graphic arts industry. The available resins, unsaturated polyesters with insufficient weather ability, have not been suitable for exterior applications. Moreover, the cross-linking density of the radiation curing systems is strongly dependent on the distance between light source and surface what results in varying surface properties when applied to three dimensional work pieces. The radiation intensity [mW/cm^2] diminishes proportionally with the distance from the UV lamp. Thus also the UV dose [mJ/cm^2] available for the activation of the photo initiator is reduced at constant process speed. Oxygen inhibition in the outer coating interface becomes more important at lower radiation doses. Fewer radicals can be formed to start the creation of molecule chains and the probability of chain termination by free oxygen radicals increases. Shadow areas that are not reached by UV light are not hardened, post cure doesn't occur.

The expected advantages, higher productivity, shorter dryers, better chemical and scratch resistance and all that at higher solids and lower emissions motivated researchers and developers at raw material and coatings producers to work on solutions for these problems. Distance dependency is faced by UV lamps guided by robots, by higher double bond concentration of the resins used and by higher levels of and further developed photo initiators. These measures have technical and economic limits. The accessibility of surface areas by robots is limited by the lamp geometry; area specific irradiation is not always possible.

Higher double bond density or higher levels of photo initiator may lead to brittle films up to cracking at distance set point. Hardening in the absence of oxygen in inert gas atmosphere strongly pushes back the distance dependency by preventing oxygen inhibition. This allows the use of diffusing mirrors that can guide UV radiation in not directly illuminated areas. Carbon dioxide, nitrogen or argon serves as inert gas.

Curing in shadow areas is also feasible by so-called dual cure systems. Next to the radical polymerization such systems have a second curing mechanism. As a rule this is a thermally

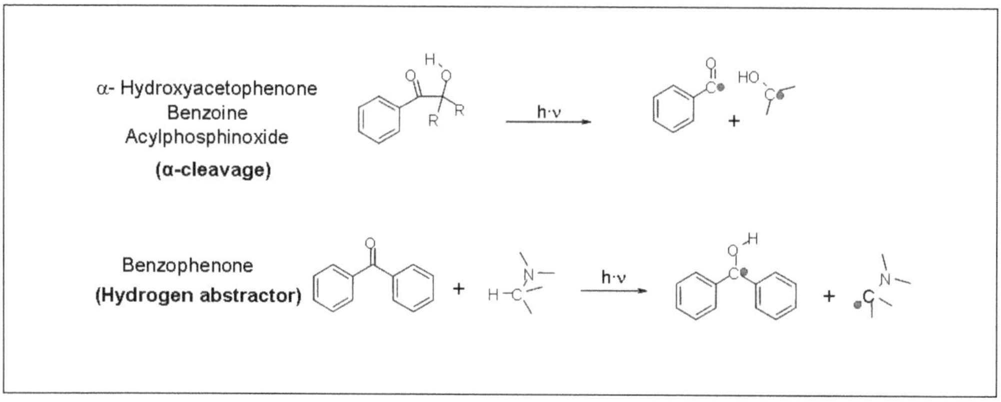

Figure 4.18: Examples of photo initiators *Source: BASF Switzerland AG, Basel, Switzerland*

induced polyurethane reaction that has the ability to post cure at ambient temperature. With dual cure systems the not directly illuminated areas exhibit lower but sufficient durability of the coating.

Radiation curable acrylates are made by various relatively simple chemical conversions of the free acrylic acid or suitable acrylic acid derivates such as hydroxylethyl acrylate with reactive base resins obtaining polyester, polyether, epoxy or urethane acrylates [5]. These acrylates may not be confused with acrylic resins whose backbone is built by polymerization of acrylic or methacrylic acid (polyacrylics, acrylic dispersions). They do not contain reactive double bonds. Urethane acrylates, polyester acrylates or acrylic acrylates are basically suitable for the formulation of UV clear coats with good weather ability. Also the necessary flexibility for the coating of plastics can be adjusted with the novel representatives of UV curable binders. An overview regarding the most important raw materials for UV coatings is given in Figure 4.17.

As already mentioned UV curable coatings need photo initiators to start the polymerization (Figure 4.18). These are additives that decompose into free radicals under high energy UV radiation by homolytic alpha cleavage or by hydrogen abstraction. The later need so-called synergists for the creation of radicals. These include amines that provide a hydrogen atom for abstraction and radical formation [17].

It is self-evident that a dilemma needs to be resolved when UV clear coats are used for exterior applications. Chapter 4.6.1.3 addressed the necessity of UV absorbers in clear coats for exterior use. It might be assumed that it is impossible to use UV absorbers in a coating that needs photo initiator activated by UV light to form radicals that start the polymerization into a three dimensional network [12]. The match of photo initiators and UV absorbers was successful in recent years. Today we have UV clear coats available that do not stand behind traditional materials in terms of weather and light resistance.

4.6.2.4 Nano clear coats

The category of nano materials includes any material having at least one of its dimensions in the range between few nanometers and about 100 nm [18]. Such raw materials are used in coatings since long, e.g. pyrogenic silica. In recent years the special properties of nano materials were investigated and they are increasingly used in all aspects of our lives. When it comes to the formulation of coatings, nano materials can dramatically improve scratch resistance and facilitate better cleaning of polluted surfaces (lotus effect, easy to clean surfaces). Nano modification can be approached by adding nano additives or nano particles (top down) or by in-situ generation of nano structures through chemical nano technology (bottom up).

Basic materials for nano particles are often silicon or aluminium oxides. The most common approach to generate inorganic particles or nano phases within an organic matrix is to utilize sol-gel chemistry of silanes (e.g., tetraethoxy orthosilane, TEOS) [18]. A major challenge is the homogeneous distribution and stabilization of the nano particles with their huge free interface in the liquid coating but also during drying. Flocculation of the particles may lead to haze, colour shift or film defects. On the other hand the huge interface is why nano materials lead to special properties of the modified coatings (see Figure 4.19). From a coatings perspective the small length scale that separates this technology, say from "micro-technology" has two key effects: first, there is a relatively large amount of interfacial material and second, the materials are optically clear [18].

The scratch resistance of a coatings film is tested by various methods, according to the intended use: material removal by abrasive devices (Taber Abraser) is measured as well as the remaining gloss after scratching by simulated car wash (Amtec-Kistler, see Chapter 6.3) or by the use of sanding paper with defined grain on a crockmeter unit.

The distribution of nano particles in a coating layer can basically be done in two ways:

- Homogeneous distribution in the entire layer
- Enrichment of nano particles in the layer surface

While the latter method has the advantage to use the (expensive) nano particles effectively the homogeneous distribution provides permanent protection even if the outer surface is eventually removed over time.

The achievable improvement in scratch resistance, measured as residual gloss, is significant as shown in Figure 4.20.

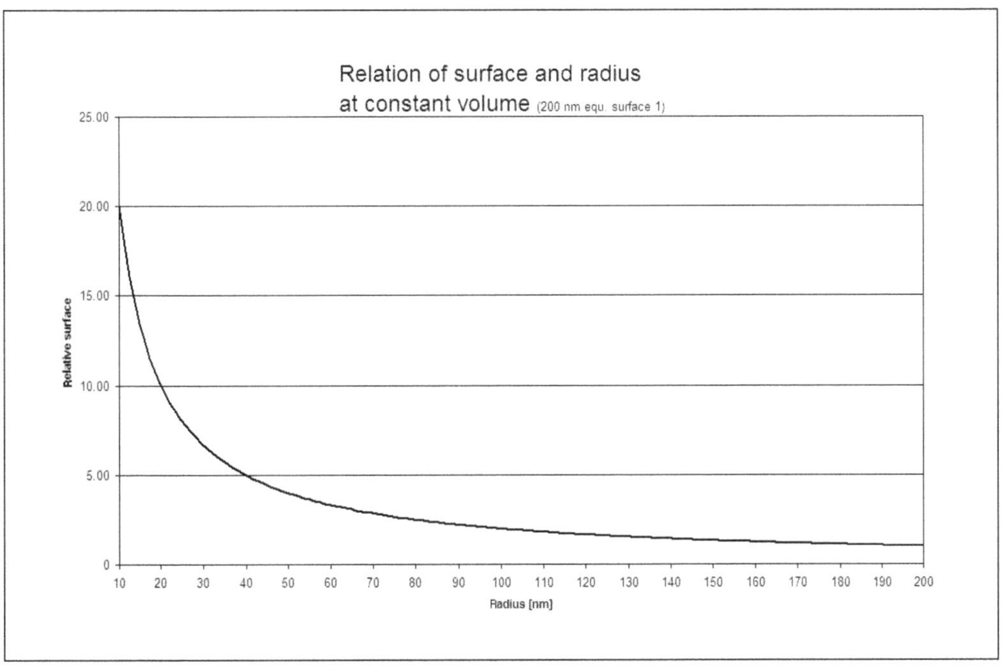

Figure 4.19: Surface area as a function of spherical particle diameter

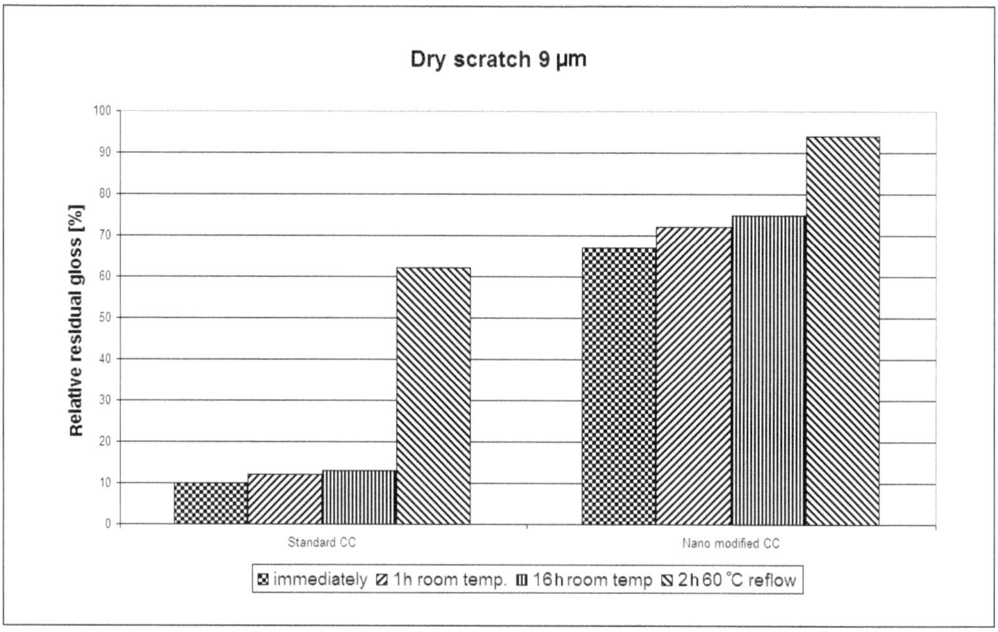

Figure 4.20: Improvement of dry scratch resistance by nano modification

Polymer glazing

Nano modified coatings find increasing application in the protection of polymeric automotive window glazing. Since many years head lamp lenses, made of polycarbonate (PC), coated with sol-gel systems, are in series production. The multiple advantages of such lenses, among others freedom of design and weight reduction, are confronted with a serious disadvantage that could be resolved by coating: Polycarbonate is sensitive to scratch and would become matt within short periods of time without protective paint.

The expansion of this application to fixed side windows and panoramic roofs took place already. Typical layer structures are shown in Figure 4.21.

Movable side windows and especially rear windows and windscreens with the action of window wipers and multiple chemical and mechanical strains need much higher durability. This is why these parts receive wear and scratch protection by chemical vapour deposition (CVD), a plasma coating. The development of liquid coatings with the necessary abrasion resistance for this application combined with the necessary preservation of the mechanical properties of the polymer and the indispensable transparency is not yet resolved. The optical characteristics of automotive windows require precision injection moulding and a coatings application technique without visible structures in the

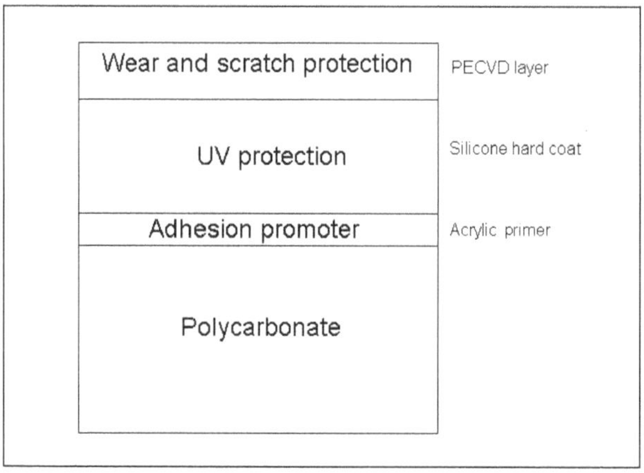

Figure 4.21: Typical layer sequence of polymer glazing[19]

final film. With liquid coatings this can be achieved by flooding. The coating is poured onto the surface; excess material can drain away and is recovered. By skilful arrangement and movement of the object while coating very smooth and homogeneous films can be produced. Beads, edges and cavities are natural limitations for this method.

4.7 Soft feel coatings

Beginning in the 1980's the automotive industry identified the car interior as a possibility to differentiate. The introduction of plastic parts allowed new geometries and designs. After a short period of time it became obvious that uncoated plastic parts show quality issues and give the vehicle interior a rather cheap ambience. Injection moulded plastic parts tend to be scratch sensitive, they often do not have the desired resistance against all kinds of chemical agents and detergents and it is difficult to produce them in uniform colour and gloss. Figure 4.22 gives an overview how the requirements for automotive interior coatings evolved over time.

The first approach for improvement was the use of low gloss topcoats as decor paints. These coatings enhanced the surface fastness and gave a much better and uniform optical impression. A striking name for these coatings was found in France: cache misère, hide the misery. The wish to ameliorate frequently touched components in their haptic behaviour lead to the development of so-called soft feel coatings. In the very beginning these were two layer solventborne systems. A primer with 15 μm film build and a soft feel topcoat having more than 80 μm were applied resulting in excellent leather like touch.

Starting in the 90's of the last century, the use of primers were eliminated and coatings were designed as a single (one) layer system. In parallel the necessary layer thickness was more

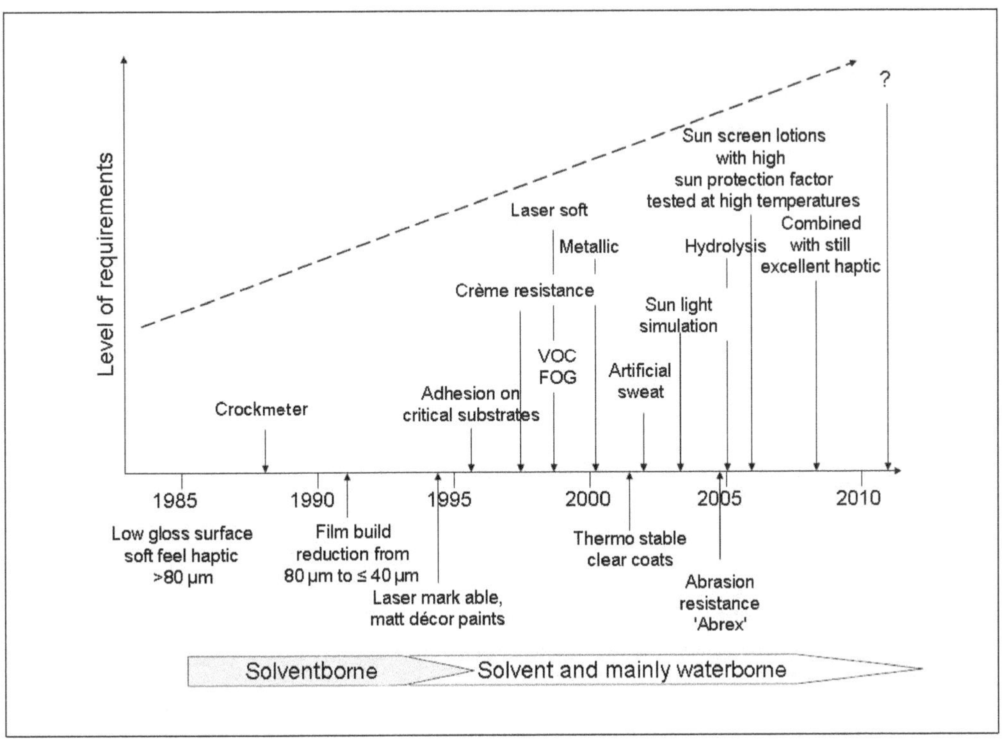

Figure 4.22: Historical development of requirement levels for interior coatings [20]

Table 4.13: Typical characteristics of soft feel coatings

	Soft feel coatings for plastic			
	Waterborne		Solventborne	
Colour/effect	dark colours	light colours	dark colours	light colours
Solids [wt.-%]	35 to 45	40 to 45	approx. 45	approx. 50
Binder-to-pigment ratio	approx. 1 : 0,2	approx. 1 : 0.53	1 : 0.05 to 1 : 0.1	approx. 1 : 0.3
Density [g/cm³]	approx. 1.05	approx. 1.1	approx. 1.03	approx. 1.1
Flow time [s] DIN4 cup	12 to 16	12 to 16	18 to 24	18 to 24
Drying /curing	minimum 30 minutes @ 80 °C		minimum 30 minutes @ 80 °C	
Film build [µm]	35 to 60	40 to 60	25 to 60	25 to 60
Gloss [units @ 60°]	1.2 to 3.0*	1.5 to 3.5*	1.1 to 2.0*	2.0 to 3.0*

** and higher*

and more reduced without substantial reduction of the haptic behaviour of the painted parts. In the mean time the car manufacturers increased their quality demands and established more and more challenging specifications. The typical characteristics of soft feel coatings are shown in Table 4.13.

Soft feel coatings need to have excellent adhesion on a variety of substrates, dominated by ABS, ABS/PC and TPO. Very good chemical resistance, hydrolysis resistance and light fastness are other key properties. The coating layers built from these systems may not emit substances that cause odour, are unhealthy to the passengers or create fallouts on the windscreen known as fogging.

Modern 2K soft feel coatings are formulated using combinations of flexible and very resistant polyurethane dispersions. They are cross-linked with aliphatic polyisocyanates, eventually hydrophilic modified to improve the incorporation in the waterborne main component. The low gloss of soft feel films is achieved by the use of matting agents. The matt surface leads to an impression of high value and supports the haptic effect. The reduction of reflections on the instrument panel helps for safer driving [16].

4.8 IMC coatings

In the in mould coating process the plastic part is painted directly in the moulding tool. The IMC coating is applied to the later visible area of the tool. Different IMC process principles are used: Painting the open cavity prior to the manufacturing of the plastic part, injection of the coating in a narrow gap in the mould, generated either by opening the tool for 100 µm to 200 µm or by injecting the paint under high pressure, leading to a slight opening of the tool. An elevated tool temperature facilitates the flow of the coating and its physical drying.

4.8.1 IMC slush

The IMC slush process is used to produce form skins for dash boards, door panels or centre consoles for the construction of automobiles. Instrument panels are often made by powder slush moulding using PVC powder. Discussions about the safe use of PVC lead to the utilization of thermoplastic polyurethane TPU. Aliphatic TPU, while light fast, is relatively expensive, so solutions for the less expensive, aromatic TPU that tends to yellowing when exposed to light, were investigated.

A possible solution is the IMC slush process. The negative cavity of the form skin is prepared using external release agents. The release agent shall allow as many production cycles as possible and needs to be compatible to the applied IMC coating. Otherwise the tool surface would not be completely wetted by the paint what would result in paint film defects. The actual IMC slush process starts with the application of the IMC coating on the 40 to 50 °C preheated cavity. It is important to keep in mind that the coating side facing the tool surface will later be the visible surface of the finished part. The coating must reproduce the usually textured surface of the mould exactly. The gloss level is solely determined by the finish of the tool surface. The use of a light sensitive, aromatic TPU to form the skin dictates the necessity to have complete hiding on all visible areas. The coatings in use are either very high solid systems (solid content >90 %) or waterborne coatings due to the fact that in most cases the production process takes place in production sites without ventilation nor air abatement. The next step is the rapid heating of the tool using sand or oil heaters. The remaining solvent or water of the coating must now evaporate completely because the cross linking starts parallel. Directly after the heating step the TPU powder is applied. It combines with the coating during gel. If residual solvent or water would be present this could provoke bubbles or other defects in the coated form skin. After cooling down to 50 °C or below the painted form skin can be released. Subsequently the skin is back foamed with polyurethane foam. Functional details such as ventilation systems or reinforcing elements are integrated in the foaming process and joined to the form skin. The applied coatings must be very flexible. No failures are accepted if the skin is sharp bent unintentionally in the final production steps. Although being very flexible the coating systems have to meet the usual requirements for automotive interior systems: abrasion resistance, chemical resistance and light fastness to name just the essentials, and cross-linked polyurethane resins are now the chemistry of choice. An important advantage of this IMC slush process is the extremely low emission measured by thermal desorption according to VDA 278.

4.8.2 SMC-IMC

When producing SMC (**s**heet **m**oulding **c**ompound) pre-pregs made from polyester resin saturated glass fabrics are formed and cured in high pressure tools. Eventually air can be entrapped between part and tool. Often a thin polymer film remains in-between the surface of the bubble and the wall of the tool [10]. When painted and cured these bubbles can open by the influence of solvents or through thermal expansion and form so-called pop ups, shown in Figure 4.23, resulting in expensive rework or rejects.

To prevent this different IMC techniques have been developed:

- **Powder IMC**: Powder coating formulated as filler (even conductive) or single layer topcoat is applied to the preheated open tool half's and pre sintered. Subsequently the part is produced as described. Through the high forming temperature and pressure the powder coating and the SMC join very well. The rather high film builds of the powder coating (approx. 100 µm) prevent from the forming of the unwanted pop ups.

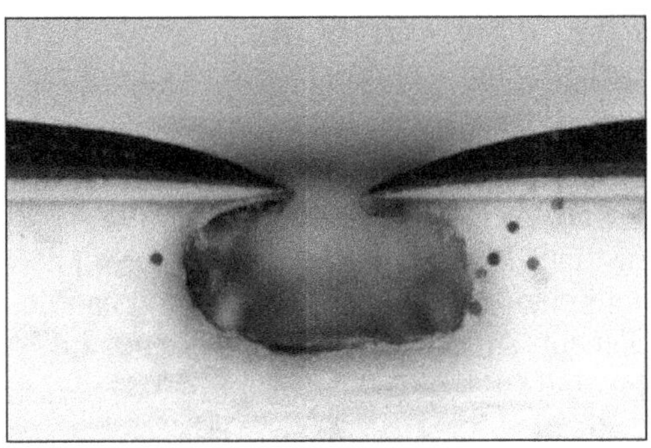

Figure 4.23: Pop up on painted SMC
Source: Karl Wörwag Lack- und Farbenfabrik GmbH & Co. KG, Stuttgart, Germany

This method is used to produce automotive parts, sanitary furniture, sports articles and cabinets for machines or electrical appliances.

- **Liquid IMC**: Liquid coatings can be used in the same way. More widespread is the procedure of opening the tool after forming the part, bring in the necessary quantity of paint and evenly spread it out over the whole surface by closing the form applying pressure. This step takes place when the part is almost form stable. The outer skin of the part is mechanically rigid enough so it is not destroyed by the paint, and the coating is cured in the mould. The coated part with primer or topcoat can then be released. A disadvantage of this method is the 5 to 20 % prolonged cycle time [21]. Another disadvantage is geometrical limitations. Tools with slide elements cannot be used. Those would disturb when opening the form.

4.8.3 High pressure IMC

Two variations of the IMC process work with high pressure when injecting the IMC coating into the tool. This can be a press tool as well as an injection mould. In one of the variants the coating is applied to the mould a certain time after the injection moulding of the part using pressure higher than the internal mould pressure. Thus the coating spreads over the entire surface of the part. The coating is formulated very reactive and must be applied using a reaction injection head. Typically, cure times will run from about 15 to about 100 seconds at mould temperatures of about 149 °C (300 °F) [21]. A prolonged cycle time has to be expected but it will be much shorter than in the liquid IMC process because it is not necessary to open and close the mould.

The second variant works with two cavities: The part is moulded in the injection or reaction mould as usual. After the standard cycle time the mould is opened and the part with the back side of the mould is brought to a second cavity in the tool where a highly reactive coating is injected and cured [22].

4.9 Painted films

The classical way to produce painted plastic parts is the production of the part itself in its final geometry with subsequent painting. The painted film technology reverses this approach. In principle decorative films for the refinement of smaller plastic parts by **i**n **m**ould **d**ecoration (IMD) are known as printed or painted films.

The IMD process combines injection moulding and hot stamping in one step. The coated or printed film can become integral element of the final part or the coating layer is transferred to the surface of the part while the film itself only serves as a carrier for the coating layer.

One of the drawbacks for the use of decorative films was the non-availability of films with sufficient width and the multitude of colours, exact colour match and surface properties necessary for the use in automotive serial production. The objective of the painted film technology is to produce visually uniform parts in such quality that these can be used in any position, horizontal or vertical, e.g. for car body exteriors. This new coating technology is not limited to automotive applications. It can be applied to surfaces of furniture, construction elements such as window frames, consumer electronics and more. The typical structure of a painted film is shown in Figure 4.24.

In the first step a special basecoat, non-blocking but thermo formable, is applied by doctor blade onto a mono or co-extruded film made from polycarbonate (PC), ABS/PC, polyester or polyolefin's like polypropylene (PP) as shown in Figure 4.24. If the desired colour is a metallic or pearlescent effect a second, thin basecoat layer is applied using pneumatic atomization to support good effect development. Solid colours do not need this second application. After thermal cure the film can be rolled up and stored waiting for the next process step. Blocking resistance of the basecoat layer is mandatory. The pressure [kg/cm^2] in the core of the coil is proportional to its length [m].

The applied clear coat is a UV dual cure clear coat, applied by doctor blade as well. The film forming of the clear coat first is only thermal, resulting in a superficial dry, further pro cess able layer. After the thermal cure a thin protective film is laminated on the clear coat to protect the surface from damage and dirt. The foil is then cut in sheets and stored before further processing.

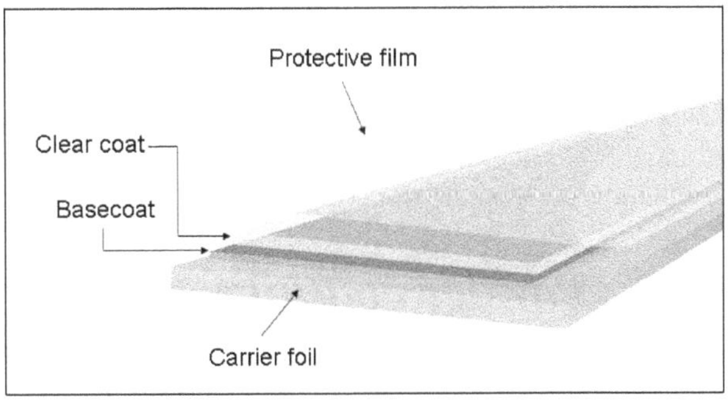

Figure 4.24: Structure of a painted film[23]

The application of the coatings via doctor blade results in a transfer efficiency of close to 100 %. Overspray is generated only in the minimal second basecoat layer application. Losses of coating material through colour change or production stops are marginal. Figure 4.25 illustrates the principle of the application by doctor blade.

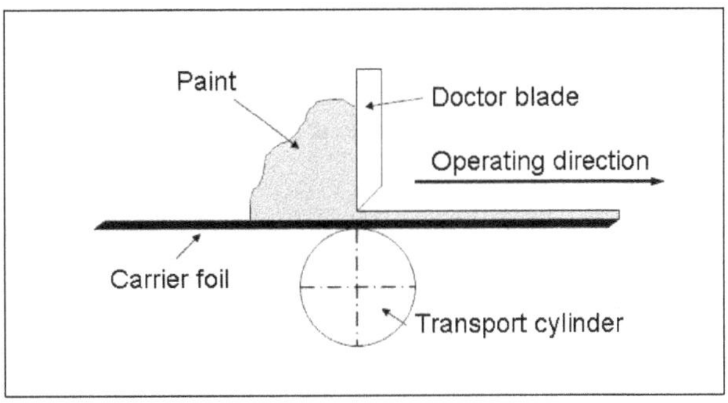

Figure 4.25: Application by doctor blade

The schematic of the manufacturing process for parts covered with painted film is shown in Figure 4.26. The steps are

- Thermo forming of the film
- UV cure of the clear coat and finally
- Back foaming with long fibre reinforced polyurethane foam or back moulding with thermoplastic polymers.

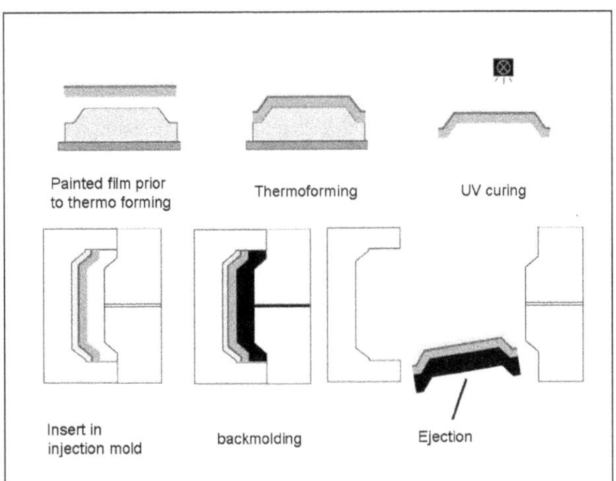

Figure 4.26: Thermo forming and back moulding[24]

The intended manufacturing processes as well as the required properties of the final part are important factors for the determination of the film material and thickness. Highly strained, horizontal components in which the film contributes to the form stability are usually made of thicker PC films (film thickness approx. 1 mm), while vertical rocker panels exposed to stone chipping are made from thin PP films, back moulded with TPO. For furniture applications or construction elements polyester films, e.g. 250 to 450 μm thick, are in use.

The UV dual cure clear coat awards the part excellent chemical resistance. The scratch resistance comes close to nano modified clear coats. The painted film technology has proven to meet the specifications of automotive OEMs (**o**riginal **e**quipment **m**anufacturers) as well as the ones for furniture or construction elements.

Paint lines belong to the most expensive elements of OEM or OES (**o**riginal **e**quipment **s**upplier) assembly lines. In consequence painting costs of a plastic part may stand for 30 to 50 % of its total cost. This share could be significantly reduced if plastic parts would be finished with painted films [25].

Recent developments are UV cured, highly flexible painted films for the application on window profiles in an extrusion process. The application of such films has just started and faces many new opportunities.

4.10 Powder coatings

The use of powder coatings for the painting of plastics today is limited mainly to the powder IMC technology (see Chapter 4.8.2). Powder coatings require minimum temperatures for film forming and curing. These are actually around 140 °C. Thus most thermoplastic polymers cannot be used as substrates for powder coatings. Actual developments give hope that the temperature limits could be reduced to 120 °C. This would allow using certain TPOs. A big challenge is the shelf life of such powder coatings. Premise for practical shelf life is the observance of a cooled logistic chain to prevent powder coatings agglomerations. Besides that the reduction of the melting temperature results in worse levelling.

UV radiation curable powder coatings are under evaluation. They also need minimum temperatures to melt as a prerequisite for film forming and levelling.

Powder coatings need conductive substrates for application. Approaches to make the plastic material itself conductive have not found their way to industrial use. The available dry film conductive liquid primers (see Chapter 4.3) could be a solution. Their compatibility with powder coatings needs to be proven. Knowing this possible powder coating applications would be powder topcoats or powder clear coats that could partially replace their liquid coating counter parts.

4.11 References

[1] Eastman Publication GN-411-C, 2004, p.5

[2] Bayer Materialscience, formulation guideline RR 4756, 2008

[3] EP 1226218B1, p.15

[4] Bayer Materialscience, formulation guideline RR 4842, 2009

[5] Th. Brock, M. Groteklaes, P. Mischke: European Coatings Handbook, 2nd Edition, Vincentz Network, Hanover, 2010, p.61 and p. 207

[6] Company information Eckart Gmbh, 2006

[7] P. Wißling et al.: Metallic Effect Pigments, Vincentz Network, Hanover, 2006, p. 24, 80 and 85

[8] Nuplex Resins bv, REC 07008, 2007

[9] G. Pfaff: Special Effect Pigments, 2nd revised Edition, Vincentz Network, Hanover, 2008, p. 16

[10] A. Huber et al, 1st German Automotive Circle, 24th Session, November 2007, Bad Nauheim

[11] G. Meichsner, T. Mezger, J. Schröder: Lackeigenschaften messen und steuern, Vincentz Network, Hannover 2003

[12] A. Valet: Lichtschutzmittel für Lacke, Vincentz, Hannover, 1996, p. 21 and 62

[13] U. Poth: Automotive Coatings Formulation, Vincentz Network, Hanover, 2008, p.187; 194 and 195

[14] Bayer Materialscience, Formulation guide line RR 4822, 2009

[15] W. Heilen et al.: Additives for Waterborne Coatings, Vincentz Network, Hanover 2009

[16] U. Meier-Westhues: Polyurethanes, Vincentz, Hanover, 2008, p. 191, 193,

[17] A. Goldschmidt, H.-J. Streitberger: BASF Handbook of Coating Technology, Vincentz, Hanover, 2007, p. 184 and 470

[18] R. Fernando: JCT CoatingsTech, May 2004, p. 33 und 36

[19] After M. Navroth: Der Weg zur Automobilverscheibung aus Polycarbonat, EUROFORUM-Veranstaltung Einsatz von Glas und Polycarbonat im Auto, Köln 2005

[20] According to S. Jacob from Automotive Paints and Coatings, Wiley, p. 342

[21] EP 0197496B1, p.2 and 6

[22] Bayer Materialscience, company information Innovationsperspektive 2006, p.6 to 13

[23] According to H. Warta, Lackfolientechnik, 11th Automotive Circle Spezial Fachtagung, Bad Nauheim, 2007)

[24] According to H. Warta, Lackfolientechnik, 11th Automotive Circle Spezial Fachtagung, Bad Nauheim, 2007

[25] DaimlerChrysler Hightech Report 1/2005, p.13

5 *Plastic coatings application*

The application methods used to paint plastic parts are as diverse as the type of parts made from plastic. Nevertheless the most important ones might be the methods using atomization of the coating material. Processes like flooding of polymer glazing or head lamp lenses (see Chapter 4.6.2.4), roller coating for flat surfaces and films or the methods to paint films with doctor blade applicators (see Chapter 4.9) are, like printing, not examined in this chapter.

5.1 *Methods of atomization*

In principal liquids can be atomized by pneumatic, hydraulic, electrostatic or rotational forces. These methods can be combined within some limits. The pneumatic high pressure spray with compressed air of 3 to 6 bar uses the kinetic energy of the atomizer air that leaves the spray gun nozzle with high velocity to atomize the paint. This method was for a long period of time first choice for plastic coatings because it resulted in excellent surface qualities even on parts with complex shapes. The most important disadvantage of this method is its low transfer efficiency, ranging between 20 to 40 %. To overcome this drawback the HVLP (**h**igh **v**olume **l**ow **p**ressure) guns were developed in the 1980s. The benefit of 5 to 30 % better transfer efficiency is paid with lower atomization quality what may lead to inacceptable colour deviations when applied for metallic or pearlescent effect coatings [1]. By adapting the spray gun settings this effect can be corrected what often results in spraying conditions (and transfer efficiencies) similar to classic high pressure atomization.

Airless atomization

The hydraulic atomization, well known as airless atomization (atomization without air support) makes use of the pressure difference (up to 200 bar) of the coating leaving a fine nozzle. Excellent square meter capacity paired with low overspray tendency and good transfer efficiency is achieved. But the feasible surface quality is often inappropriate for decorative applications. The high material flow makes exact film build control critical and well trained personal is necessary to carry out the coatings operations. An approach for improvement is the so-called air mix procedure that combines the airless method with compressed air support.

Hot spray

All spray methods mentioned so far can be done with coatings materials heated up to 40 to 60 °C, the so-called hot spray. The higher material temperature lowers the viscosity. The material can be applied at 5 to 10 % higher solid content, resulting in better transfer efficiency. Levelling and drying can be influenced in a positive way. When used with 2K coatings the accelerated reaction through higher material temperature needs to be taken into consideration.

Wilke/Ortmeier: Coatings for Plastics

ISBN: 978-3-86630-859-6

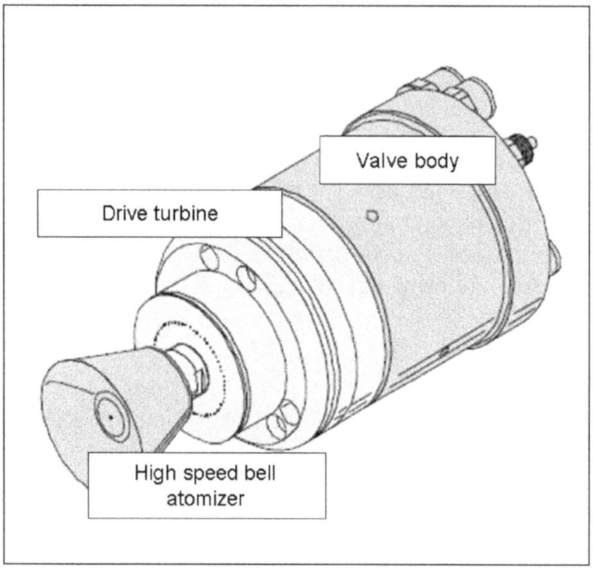

Figure 5.1: High speed bell atomizer

Source: Th. Scholz, LacTec Paining Systems, Rodgau, Germany

Electrostatic atomization

Coatings are applied using electrostatic atomization since approx. 1950. On fixed or slowly rotating atomizers the physics of electrostatic fields bring about the formation of paint droplets. Through the simultaneous transport of electrically charged paint along the electrical field lines to the grounded work piece and the inner repulsion of the paint jet, charged with similar polarity, which is formed by this transportation process, the paint jet falls apart in single droplets. The very good transfer efficiency and the wrap stand for the electrostatic atomization, the limited square meter capacity and low film builds in Faraday's cages are drawbacks. Pure electrostatic atomization requires substrates with metal conductivity so it is usually not used for plastic coatings applications.

Pneumatic atomization

Pneumatic coating methods can be supported electrostatically. One uses the material transport of charged paint droplets along the field lines what leads to better transfer efficiency and the so-called wrap around the work piece (see Chapter 4.3: Primers, Figure 4.4: Principle of the electrostatic coating method). For this method conductive substrates are necessary as well. The conductivity can be achieved by the use of conductive primers as explained in Chapter. 4.3.

High speed bell atomizer

The equipment manufacturers have further increased operating and transfer efficiencies by increasing brush speed (up to 1000 mm/s), increased fluid flows (>600 ml/min) and developed more reliable air bearing turbines so that transfer efficiency can be optimized. These advances made during the 1990s have now found wide spread use in applying plastic coatings as well.

Fast rotating bells, up to 60,000 min^{-1} atomize the coating material by centrifugal force. The coating gets charged either directly through contact with the bell that is connected to high voltage or by external charging through external electrodes and guided along the electrical field lines to the earthed coating object. The transfer efficiency can be as high as 70 % and more combined with excellent surface capacity. Figure 5.2 illustrates the principle of direct charge.

Trying to shift the process limits to further and further increased productivity is obvious. Inferior surface quality caused by too high material flow related to the available atomizing capability and reduced yield caused by too high brush speeds often are possible consequences. In the first scenario the atomizing power is not sufficient to form the desired fine and even droplet spectra. In the second case one can imagine that the electrostatic field

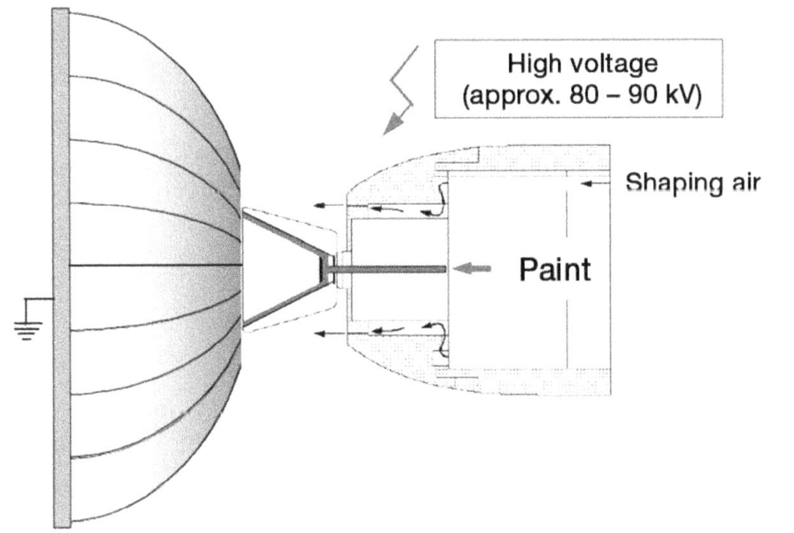

Figure 5.2: Contact charging *Source: Th. Scholz, LacTec Paining Systems, Rodgau, Germany*

moves too fast to be able to guide the cloud of coating droplets in an effective way towards the substrate.

Electrostatically assisted high speed rotational atomizers

The use of waterborne coatings is very advantageous when using electrostatically assisted high speed bell application for the coating of plastics. Waterborne coatings in liquid phase are conductive and can conduct the surface charge via the developing coating film to the earthing point known as dynamic earthing. The necessary potential separator needs to be considered; otherwise high voltage can leak back from the applicator to the paint supply endangering employees and equipment. The principle to derive the electrical charge by a conductive liquid coating film was recently successfully transferred to solventborne materials, e.g. conductive clear coats for primer-free application.

High speed atomization without electrostatic support

The newest developments in the technology of atomizers are high speed bells without electrostatic support. They use the shaping air to form the spray and to direct the material towards surface, cavities and even to penetrate recessed area impacted by Faraday Effect.

5.2 Coatings processes

The best suitable coating process will be chosen dependent on the criteria shown in Figure 4.1. Substantial topics are

- Size
- Form
- Complexity

of the parts that will be painted.

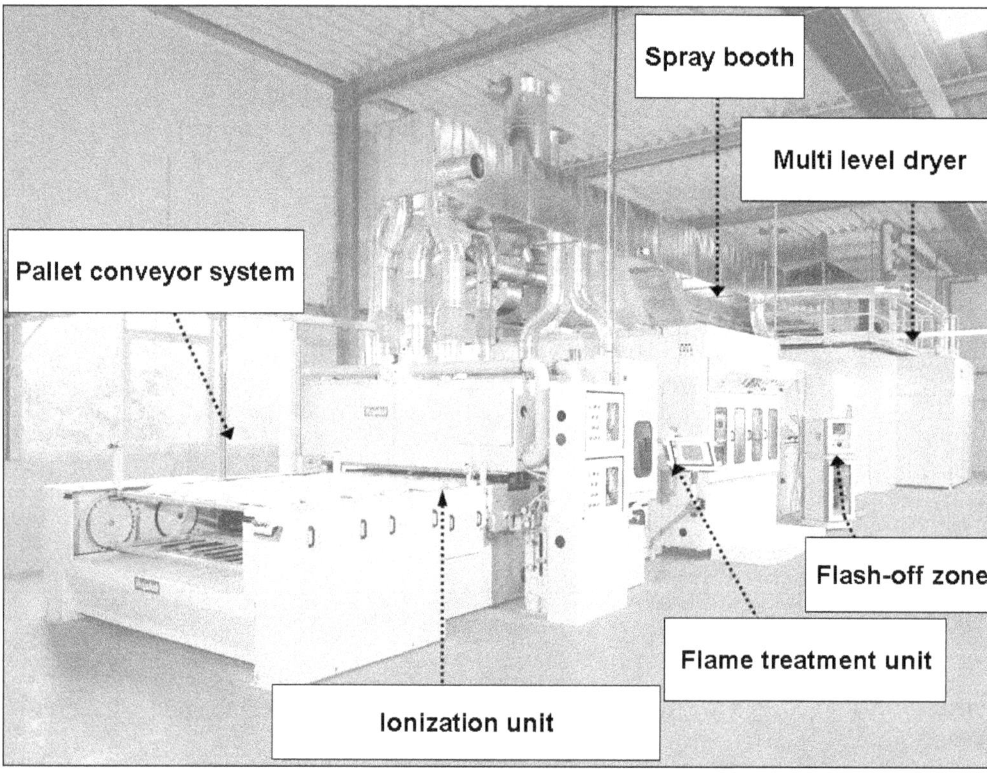

Figure 5.3: Flat bed coating machine *Source: Venjakob Maschinenbau GmbH & Co. KG, Rheda-Wiedenbrück, Germany*

The number of similar parts decides about the possible atomization. The requirements in quality and optics are the fundamentals to decide the paint system, necessary number of paint booths, dryers, inspection and additional work zones for sanding, masking or polishing. For rather small scale articles flat bed or round table coating machines with pneumatic gun application, eventually with electrostatic support, will be first choice. Special series will be best carried out by hand spray. Large parts for the original equipment supply to the automotive industry, e.g. fascias, are for the most part painted in fully automated paint lines by robots equipped with high speed bells.

5.2.1 Flat bed coating machines

Flat bed coating machines transport the work pieces on flat conveyors to the different stations of application (see Figure 5.3). Part feeding can be followed by a pre-treatment zone with flame or corona treatment. In the spray booths the coating is applied by spray guns, moving right angled to the conveyor orientation. Flat or two dimensional formed parts can be painted in one passage. Adjusting a flat bed coater to good spray results needs lot of experience. Positioning of the parts on the conveyor needs to be optimized for efficiency (densest possible packing to minimize overspray) and surface quality (avoid shielding from part to part). The number of guns in use and the relative velocity of the gun carrier against the conveyor need to be synchronized. The guns need proper orientation to cover all surface areas accordingly.

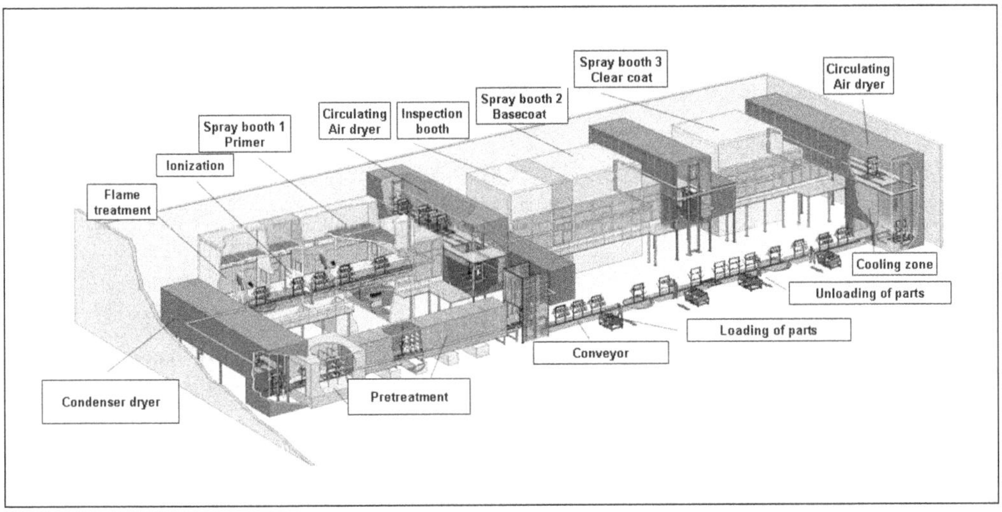

Figure 5.4: Paint line for plastic fascia *Source: Eisenmann, Böblingen, Germany*

The spray booth is followed by a flash-off zone and the curing oven as usual. For multi layer coatings several passes through the paint device or installations with the necessary numbers of booths are possible solutions.

5.2.2 *Large scale paint lines*

Paint lines for large series of big plastic parts such like fascias are not much different in complexity and techniques applied compared to automotive body paint lines. Each automobile needs two bumpers, rocker panels, probably body side moldings, mirror housings, sometimes fenders or trunk covers and other body parts that nowadays are predominantly painted in body colour. Efforts to improve colour harmony between supply parts and car body lead to a convergence in application processes. Robots with high speed bells are today's paradigm in automotive OEM and OES paint lines.

Figure 5.4 illustrates the architecture of a paint line for fascias. Parts feed onto supports, so-called conveyor skids, usually built as under floor conveyors to avoid dirt contamination of the fresh paint, is followed by automatic parts cleaning in a multistage power wash installation. In subsequent treatment zones the parts are sprayed with detergents that are finally rinsed with demineralized water to avoid residues of mineral salt or cleaning agents. This final rinse must be very effective. Such residues can lead to visible coating defects or form bubbles and lead to adhesion failures in later use of the parts. The superficially remaining water is removed in blow zones and the adhesive water dryer. Problematic are part cavities that carry water. Often the adhesive water dryer is not sufficient to remove this amount of water that can drop on underlying parts and surfaces and form surface defects and rejects. Costly manual blow out of these areas often is the only counter measure.

Pre-treatment is followed by an activation step, in most cases flame treatment with an oxidizing gas flame to increase the surface tension by incorporating polar groups into the polymer surface. Primer booth, basecoat and clear coat booths with the necessary flash-off zones and dryers succeed.

A typical process sequence is shown in Table 5.1. Variations are possible and often made at any point. Solventborne coatings allow so-called wet-on-wet processes partly or completely

Table 5.1: Typical process times in paint lines for plastic parts

Coating process	Standard conditions for waterborne	Minimum conditions for solventborne
Loading of parts	5 minutes	5 minutes
Power wash and blow zone	10 minutes	
Water dryer	20 minutes	
Cooling zone	10 minutes	
Flame treatment	5 minutes	5 minutes
Cooling zone	5 minutes	5 minutes
Primer application	5 minutes	5 minutes
Flash off	10 minutes	5 minutes
Curing	20 minutes	
Cooling zone	5 minutes	
Basecoat application, 2 times	10 minutes	10 minutes
Flash-off	10 minutes	5 minutes
Curing	15 minutes	
Cooling zone	5 minutes	
Clear coat application	5 minutes	5 minutes
Flash-off	10 minutes	5 minutes
Curing	45 minutes	30 minutes
Cooling zone and unloading	15 minutes	10 minutes
Total	210 minutes	90 minutes

without curing steps between the individual coating layers. Such processes require extremely fast physically drying coating formulations. The application window, that is the bandwidth of possible process parameters to achieve a good paint result, is therefore relatively narrow.

The plastic parts are finally examined by a surface inspection. Smaller defects are reworked by purposely sanding and polishing. Larger defects (dirt inclusions, sags, thin coating) are often recoated passing the paint line again after sanding out of the defect. The economic viability of this proceeding is under constant discussion as the success is usually not as good as the normal first run rate[1].

5.3 References

[1] Th. Brock, M. Groteklaes, P. Mischke: European Coatings Handbook, 2nd Edition, Vincentz Network, Hanover, 2007, p. 290

1 *First run: Painted part with a surface quality meeting all requirements without further touch.*

6 Coatings on Plastics

6.1 General aspects

The technical properties of industrially coated plastic parts are investigated with specified test methods and compared with the demands of specifications. These demands can be classified into basic characteristics and application-specific properties. The latter depend on the kind of object to be coated, whether it represents an injection-moulded part or for instance an extruded foil or plate. Within these two groups of coated objects, the diversity of specifications might be large and accentuated very differently. Very often certain properties are paramount. Table 6.1 presents examples of the heterogeneity of specific demands.

In order to gain sustainable acceptance, plastic coatings have to fulfil some basic requirements, which are of central importance:

- Adhesion
- Resistance to mechanical stress
- Chemical resistance
- Optical properties

Robust coatings do have to fulfil these demands even when they are brought below or over the recommended cure temperature, or if the film thickness is out of the suppliers' specification. The key expression of robustness is a large process window, representing the maximum tolerance range of all possible application and curing parameters in order to match the specification. The influencing parameters of a coated plastic part are the coating process and material as well as the plastic material and processing.

All these factors refer to the manufacturing process of the product, whose robustness is represented by a broad range of process parameters. The boarders of that process window should be identified and communicated. It should be added here, that the demands on coat-

Table 6.1: Specific demands of plastic coatings

Field of application	Product demand
Automotive interior	Haptic properties, chemical resistance
Glazing	Scratch resistance
Packaging	Barrier to diffusion
Vinyl flooring	Barrier to plasticizers
Windscreen wiper blades	Slip, exterior durability

Wilke/Ortmeier: Coatings for Plastics

ISBN: 978-3-86630-859-6

ing properties not only are imposed for the initial state, but also for different aging times and conditions, such as climate change, condensation, as well as mechanical and chemical stress. It is generally required, that the coatings adhesion, protective functions and aesthetic character remain completely unchanged after being exposed to weather, mechanical impact and chemical attack. If these stress factors are integrated into the description of the product's robustness, then the resulting window not only shows the characteristics as a function of process parameters, but also as a function of stress factors, so it should be three-dimensional.

Before beginning with coatings tests, the film thickness should be checked carefully, as coatings properties strongly depend on it.

There are direct and indirect methods for the inspection of coatings thickness.

- Indirect methods
  - Simultaneous painting of metallic platelets, which are used for the electromagnetic inductive determination of the film thickness
- Direct methods
  - Microscopy (destructive)
  - Thermal waves (non destructive)
  - Ultrasonics (non destructive)

The requirements of plastic coatings, concerning chemical resistance and optical properties do not principally differ much from those of coatings for metallic substrates. The most important aspects are described in Chapter 4, with basecoats and clear coats. The demands on automotive application, which is the most important field of application for plastic coatings, and the necessary test methods are described profoundly in the literature [1]. Therefore a complete description of coating properties and their test methods shall lag behind some plastic-specific issues like adhesion and mechanical properties. The following pages at first briefly describe the characteristics belonging to those issues and then some practical test methods including, if existing, scientific investigation methods.

6.2 Adhesion

6.2.1 Modes of failure

Adhesion is a characteristic, which with almost no exception (special temporarily applied films) is a requirement of every coating. Many other coating functions are connected to adhesion. This fact also is valid for plastic coatings and for plastic substrates with low wettability, where often obtaining adequate adhesion is hard to achieve.

One might ask at this point, what it means if coating adhesion on plastic substrates does not take place, or how the separation of a coating from a plastic substrate looks like. From experience, there can be three types of adhesion failures with coated plastic parts: (Figure 6.1).

- **Adhesive failure** is characterized by separation of the coating from the substrate or from a coating layer underneath, whereby the location of separation is the interface between both materials.
- **Cohesive failure of a coating** means, that the separation of the coating from the substrate is located within the coating layer. One might presume, that in this case the cohesive forces in the coating layer are weaker than those in the substrate and also

weaker than the adhesive forces between coating and substrate. The cohesive properties of a coating in principle can be investigated by tensile tests on free films (see Chapter 2.2.1).

- **Cohesive failure** can also take place underneath the coating, **within the substrate**. In this case the detached coating layer is carrying substrate material on its backside, giving incidence that the substrates cohesive forces are weaker than those of the coating and the adhesive forces.

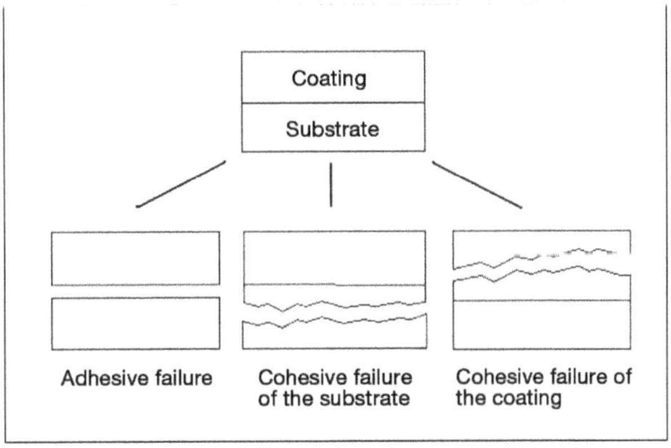

Figure 6.1: Adhesive and cohesive failure of a coating on plastic substrates

It is possible to identify the location of cohesive failure by microscopy. The backside of detached coating layers can be examined by Scanning Electron Microscopy (SEM) and Energy-Dispersive X-ray analysis (EDX) to identify possibly attaching substrate material by its elemental composition (Figure 6.2).

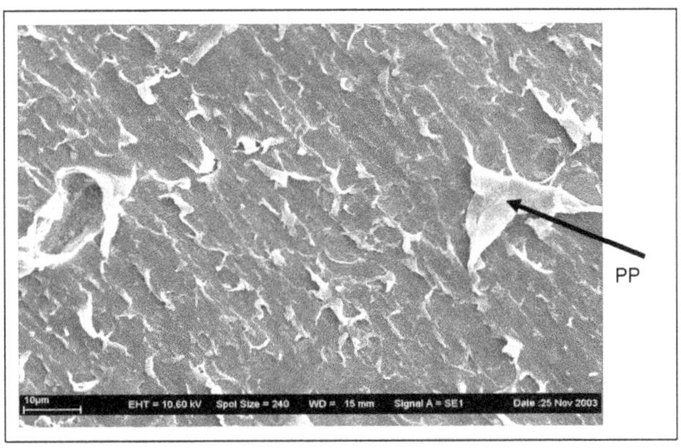

Figure 6.2: SEM of substrate residues (PP) on the backside of a coating layer

By examination of the other layer side, one might also find residues of detached coating on the substrate by this method (Figure 6.3).

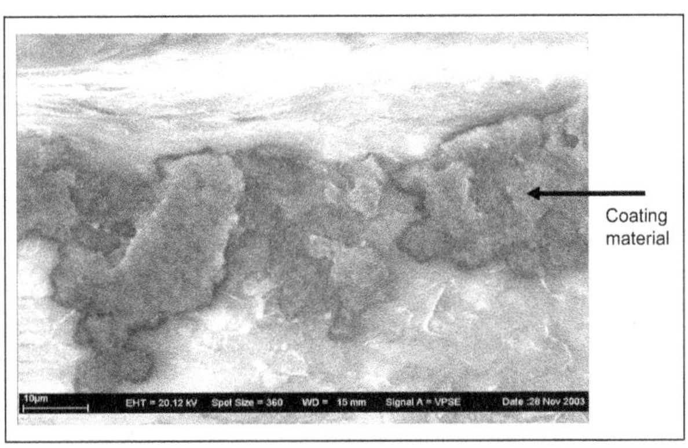

Figure 6.3: SEM of coating residues on TPO

Both kinds of adhesion failure can also occur side by side. This is only one possible reason, why a root cause analysis with adhesion failure may become very difficult. Another problem is that often coatings detach from plastic substrates on different locations, seemingly by coincidence, and without any systematic pattern, even not on every moulded part of one series. Very often this kind of vagrant adhesion failure cannot be reproduced in the laboratory of the coating supplier, exclusively taking place on original parts coated at the customers paint line, treated with extreme loads, like the water steam jet applied on scribed coating surfaces (Figure 6.4).

Basically it is fairly safe to say, that adhesion of a coating on plastic substrates depends on the quality of the coating material

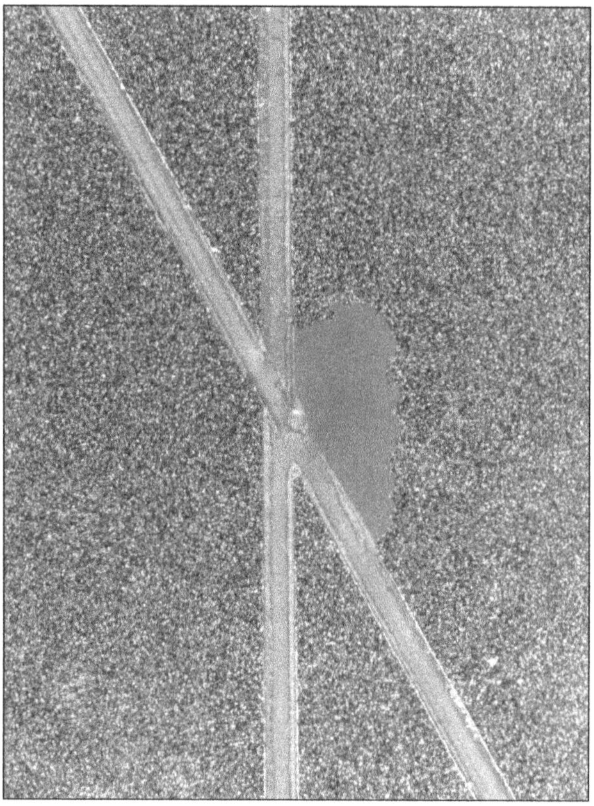

Figure 6.4: Coating adhesion failure after steam jet test (from [1])

and its application, on the status of the substrate material and surface after being processed to a moulded part.

6.2.2 *Adhesion theory and -mechanism*

The sum of all molecular interactions between the coating and the substrate as well as within both materials results in an integral material characteristic of the whole polymer laminate. This property is perceived and evaluated in tests of coated plastic parts and referred to as adhesion. It would be desirable to understand to which degree which kind of interactive force is responsible for the adhesion in every case of coated plastic parts. From a physical point of view this would mean the attempt to relate the results of adhesion tests, from perfect attachment to complete failure, to each physical cause of those results. By trying to follow this approach, one is soon confronted with a first problem. Defining and using adhesion in the common way means to work with a double sense of adhesion. On one hand the term adhesion is used as the general expression of the phenomenon. On the other hand the different kinds of adhesion failure, as described above, do rather suggest preferring a term like composite adhesion or either cohesion, as only one of three failure mechanisms relate to pure adhesive interactions. This counts at least for coatings on polymer and other non metallic substrates, as on metal one should not expect some kind of substrate related delamination . In the following briefs, the theories and mechanisms of adhesion shall be presented briefly, except the **weak-boundary layer concept**, which already has been presented in Chapter 3.

From another standpoint, thermodynamic adhesion is described as an interfacial tension (or interfacial energy) in between the coating and the substrate. It is determined as work of adhesion in Nm/m^2 (=N/m), which is afforded to detach the coating from the substrate [2].

Modern adhesion research is occupied with the mechanisms, which on a molecular level cause adhesion. Wetting of the substrate is considered as one basic requisite, but not the only root cause of adhesion. Depending on the kind of literary reference there are four to five different basic mechanisms which should be theoretically understood [3-5].

Mechanical theory

This mechanism of adhesion is based on the assumption, that a high surface roughness promotes the mechanical anchoring of the coating material, as during film formation it penetrates into pores, holes, cracks and voids before getting solid [6] (Figure 6.5). For example the adhesion of metal layers on ABS plastic could be related to this kind of mechanism. The ABS does have to be pretreated in a way that a surface structure is created that allows anchoring of the galvanic metal layer. The target of such a measure is the increase of the surface area getting in order to enhance contact with the coating material. A common way to increase surface area of plastic parts is sanding, as it is performed in the serial production of many coated fibre reinforced components for automotive and aircraft vehicles.

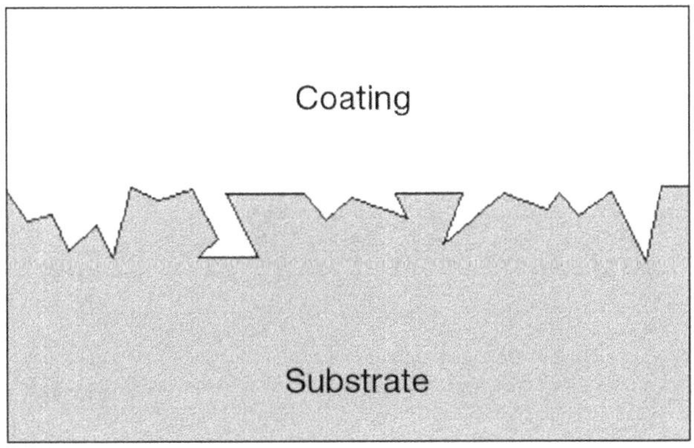

Figure 6.5: Adhesion mechanisms: mechanical anchoring

Chemical bonding theory of adhesion

The formation of covalent bonds in an interface between coating and substrate is possible and can take place easily with some curable coatings. This type of bonding is considered as the strongest and longest lasting one. But it affords the existence of matching reactive groups on the substrate surface with those in the coating material. Covalent bonds are difficult to prove, although this has already been achieved in the past. Some plastic substrates contain diverse functional groups, which in principal can react with certain coating materials. Organosilanes are used as adhesion promoters for glass fibres in reinforced plastics. The reaction between silanole-groups of both partners provides the fibre with a surface being more wettable by organic coating binders.

Coatings with reactive groups like carboxyl- and hydroxyl-groups do often adhere well on surfaces which are equipped with the same or similar polar groups. So acrylic melamine topcoats often show a good adhesion on primers composed of polyester-melamine. It is assumed that this is due to reactions between remaining hydroxyl-groups of the primer

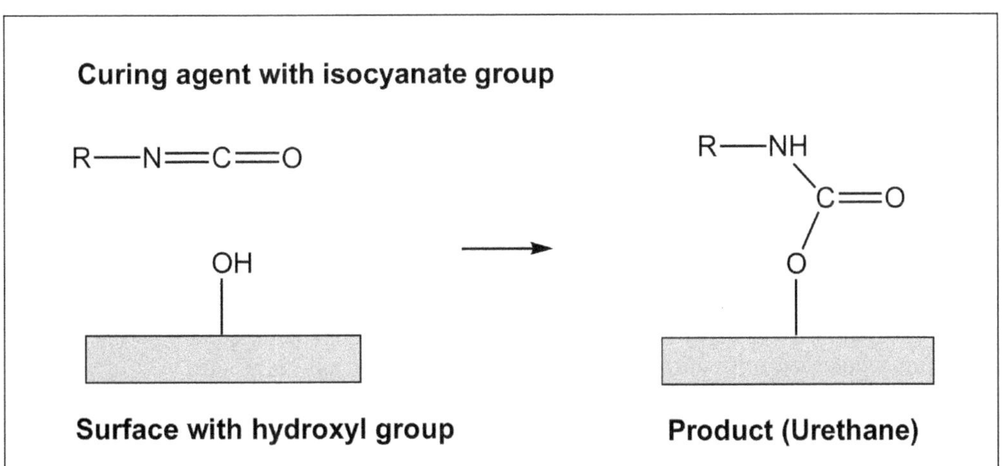

Figure 6.6: Adhesion mechanisms: chemical bonding, reaction of an isocyanate cross-linker with hydroxyl groups of the substrates polymer surface

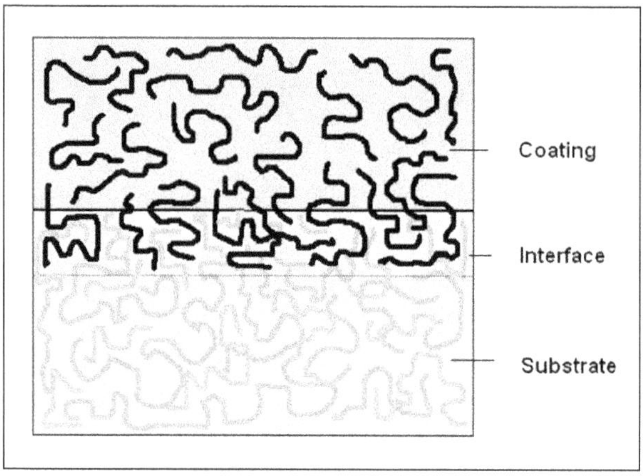

Figure 6.7: Adhesions mechanism: diffusion theory

with residual melamine resin from the topcoat. After overbaking the primer, adhesion becomes worse. Another kind of bond formation, which appears to be quite obvious and probable, is the reaction between isocyanate groups of 2K polyurethane coatings and hydroxyl groups of polyester-, cellulose- and oxidized polyolefin substrates (Figure 6.6).

Electrostatic theory

The polymer chains of coating and substrate have a more or less high polarity. Permanent and induced dipoles at the surface of both phases orientate in such a way, that they turn countercharges to each other and thus form electrical double layers. Other kind of electrostatic adhesive forces are dispersive interactions. All those kinds of bonds can be labelled as physical. Bond strength decreases with the sixth or seventh power of the distance. That is why they usually act only over small distances, roughly up to 0.5 nm.

Diffusion theory

Basically the polymer chains of substrate and coating material can cross the interface to penetrate into the respectively opposite phase. *Voyutzki* and others have proposed, that adherence of a polymer composite comprising a substrate and coating is caused by interdiffusion [7-9]. Any kind of local polymer chain transfer into the opposite composite phase implies proper wetting at first. Furthermore the degree of inter-diffusion should strongly depend of material properties and curing conditions. Similar polymer properties, for example an equal coefficient of diffusion, should ease inter-diffusion. It could be demonstrated by theory that chain segments of different polymers form a boundary layer of 1 to 100 nm thickness, as presented in Figure 6.7 [3, 4].

Which proportion the above presented mechanisms each have, is not known. Moreover their proportions depend of the kind of substrate and coating material coming into contact with each other. Finally there is no valid indication about how a surface for optimal coating adhesion does have to look like in order to gain maximum adhesion.

6.2.3 Adhesive strength

The molecular forces, causing adhesion of a coating on a substrate, are not accessible for experimental investigation. But adhesive interactions together with cohesive forces can be perceived in a macroscopic way, by their joint effect on practical coating adhesion tests results. Adhesive strength of a coating on a given plastic substrate is characteristic based on this joint effect of cohesive and adhesive forces and therefore can be more precisely considered as a bond strength of the coating-substrate polymer laminate.

According to EN ISO 4618 adhesive strength of a coating is defined as the collectivity of binding forces between the coating and the substrate.

Depending on the test method used, different values for determining adhesive strength can be obtained. There are both qualitative and quantitative methods for determining this.

6.2.3.1 *Qualitative test methods*

The desire for permanent adhesion can be formulates as follows: The coating should stay attached to the substrate, under all kinds of conditions. This approach is fundamental for the following practical and qualitative adhesion test methods, which differ quite strongly in their way of execution and applied stress. Not all adhesion test methods that are common for ferrous substrates can be used for non-ferrous substrates. These are mainly the crosshatch test method according to ISO 2409, then various manually applied scratch tests with knifes as well as steam jet tests. A more precise presentation of these test methods is beyond the scope of this book, but subject of relevant standards (see below) and specialist books [1, 10].

The crosshatch test according to ISO 2409 benefits from its simplicity and suitability as a field test. Nevertheless it is a destructive method. The coating has to be scribed with a special knife in order to get a rectangular grid of notches. The distance between the notches (in mm) depends on the film build (in µm) and ranges from 1 mm (0 to 60 µm), 2 mm (60 to 120 µm) or 3 mm (120 to 250 µm). While scribing the coating it can already chip of the substrate caused by insufficient flexibility and/or adhesion. This damage picture can be enhanced by application of a specified adhesive tape and peeling it off. The degree of failure is then expressed by the number of a damage picture on a reference scale.

Scratching the intact surface or even scribing it with a knife before to applying adhesive stress, may give results with certain evidence but also limitation of reproducibility, depending mainly on the test person. Manually applied knife scratch tests afford special experience of the person performing the test, in order to receive significant results.

Steam jet tests are established methods in the field of automotive coatings. They are generally considered as decisive tests, applied after the crosshatch test has been passed in an acceptable way. A hot water steam jet applies a particular stressful load on interfaces, coating material and also substrate material. This test duplicates a vehicle being pre-cleaned prior to going through an automated car wash. The steam jet test is at best when performed in a special chamber equipped with a pressure adjustable jet. The chamber allows for adjusting time, pressure and distance from part depending on OEM specifications. This test is normally conducted when a St. Andrews cross has been scribed into the coating down to the substrate, according to ISO 3409. The steam jet is applied on to this scribe and the surface is inspected in terms of adhesive failure.

There are about 15 different steam jet test versions existing in the automotive industry. The German standard DIN 55662 is the result of efforts to achieve a harmonization of the existing different specifications. The steam jet device should allow an adjustment of parameters in the following range (Table 6.2).

Furthermore the DIN 55662 prescribes three variations of procedure, presented in Table 6.3 on page 110.

The steam jet parameters have to be controlled carefully, as their constancy decides on the reproducibility of the test results. This includes calibration of the nozzle jet's geometry using a polystyrene foam [1]. In order to achieve statistical support of the results, the steam jet test is carried out in up to 20 determinations.

Table 6.2: Steam jet parameters

Parameter	Range
Volume flow	8 to 14 l/min
Temperature	20 to 80 °C
Pressure	40 to 130 bar
Duration	10 to 300 s

All qualitative test methods have in common, that they do not deliver any physical characteristics,

Table 6.3: Variations of steam jet test procedures, according to DIN 55662

Procedure	Water temperature [°C]	Impact angle	Distance nozzle/sample [mm]	Volume flow [l/mm]	Duration [s]
A	60±2	90±2°	100±1	11,3±0,2	30±1
B	60±2	90±2°	100±1	11,3±0,2	60±1
C	60±2	90±2°	130±1	11,3±0,2	60±1

but rather an answer to the question, whether the coating is still on the substrate after it has been stressed.

This is surely not the case, if the coating

- has detached completely or partly, or
- has broken internally.

The above described practical tests cannot provide adhesion results in a quantitative way. It would be rather nice to see their reproducibility being confirmed in every case. Another interesting point is the need for understanding adhesion. There are some quantitative methods by which one may come more close to an evaluation of adhesion free from disturbing influences.

Mechanical stress, that causes adhesion failure, often acts as a combination of compression and shear. The collision of automobiles at slow speed might serve as a good example. The low mechanical stress allows plastics components to keep in shape, but the coating might detach from the substrate due to mechanical damage, and combined compression and shear stress in the collision zone. This kind of complex combined stress can be simulated in the lab with the help of special devices like the US American SLIDO apparatus that slides on the surface in a multi-axial way, squeezing the coating by simultaneous compression and shearing. Analysis of the test data allows statements about adhesive and cohesive portions of adhesion failure [11].

6.2.3.2 Quantitative test methods

While the qualitative test methods have the advantage of practicability and simplicity, the hereafter described methods are intended to obtain measured values of adhesive strength. As can be seen, this is not possible to a high extent in every case, but often only with limitations and approximations. In the following only those procedures will be presented, that have found some application and popularity in the business of plastic coatings. These are the **peel test, direct tensile** and **shearing tests** of cemented circular probes, and the **instrumented knife cutting test**, also called **hesiometer test**. They all have in common, that they use an instrumented removal of the coating and measurement of the force needed for this. The methods differ mainly in the kind of removing the coating, either by peel or pull off, and in the resultant experimental setup, with necessary apparatus and tools. There are some other methods for the examination of adhesion, which are not presented here but described elsewhere, including a special setup of the three point bending test and diverse variations of the blister test [12, 13].

Pull-off test of adhesion

This examination method is described in ISO 4624 and suitable for the estimation of the adhesive strength of a single layer coating or a multilayer coating, by measurement of the minimal tensile stress, which is necessary to detach or pull off the coating from the subsurface.

The test is prepared by cementing a circular test stamp on the coating. Figure 6.8 shows a scheme of a typical testing arrangement.

The test stamp is pulled off vertically by a tensile test machine with defined speed. The force necessary to detach the test stamp, is related to the cross sectional area, giving the adhesive tensile strength according to DIN EN 13494. This characteristic is also called the adhesion value, representing the coatings tensile stress at break in vertical position, in units: N/mm^2 (= MPa). Information about cause and location of the break has to be added separately in the test report. Cohesive failure can take place either within the substrate, the coating or the adhesive. Adhesive failure can take place between substrate and first coating layer, or between the layers (interfacial adhesive failure), between the topcoat layer and adhesive and between the adhesive and test stamp. In order to improve the delimitation of the surface area to be tested from the surrounding surface area, residual adhesive and coating material situated below can be removed by milling or cutting. The following information is important in order to obtain significant results:

- The adhesive is to be selected carefully. It has to have a higher cohesive and adhesive strength than the coating to be tested as well as a higher tensile strength than the plastic substrate.
- Substrate, test stamp and tensile test machine have to be adjusted in coaxial position.
- The tensile force has to be applied in vertical direction to the coating surface and distributed evenly over the tested surface area.
- The break has to take place within 90 seconds after the test has begun.
- Thin and flexible substrates have to be reinforced.

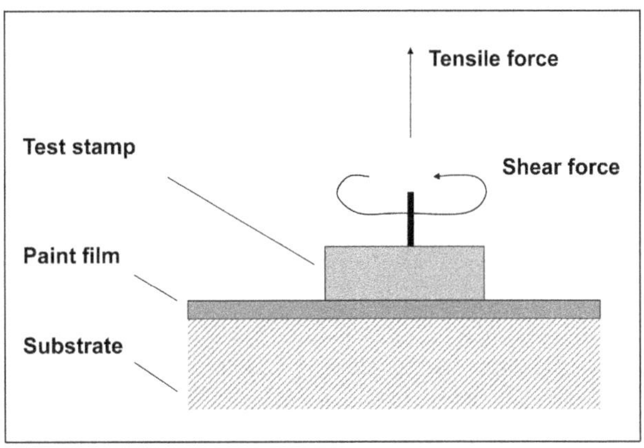

Figure 6.8: Scheme of a testing arrangement for the pull-off test according to ISO 4626

Peel test

In this test a stripe of coating film is pulled of the substrate using a tensile test machine. Important parameters to be defined are the width of the coating stripe and the angle between tensile force and coating surface (Figure 6.9).

There are several variations of this test method, which mainly differ in the kind of sample preparation. In most cases a textile fabric is imbedded into the coating, by which the deformation of the

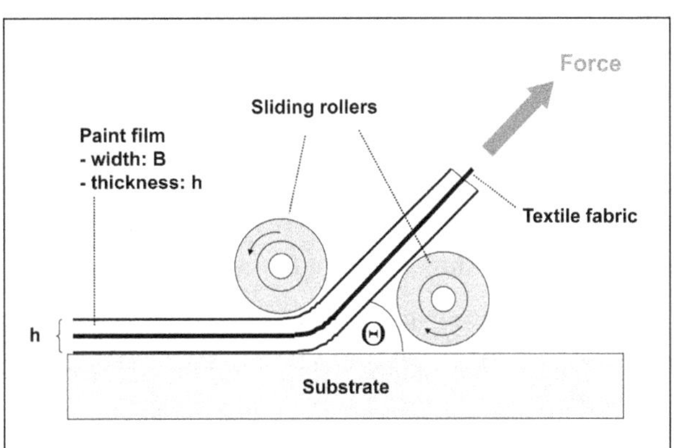

Figure 6.9: Scheme of the Peel test

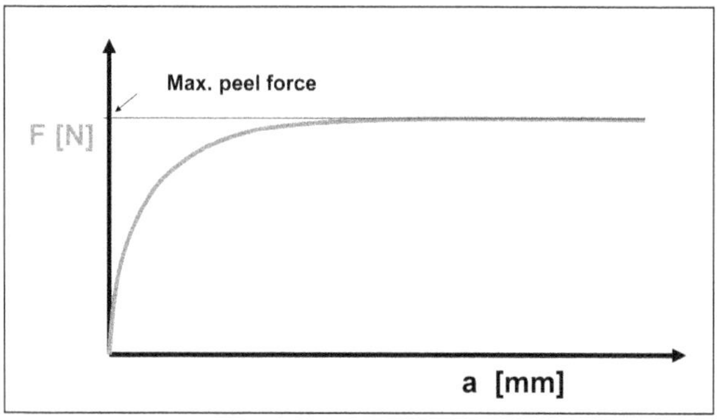

Figure 6.10: Peel test curve (simplified scheme)

coating film and resulting influence on the measured force is minimized.

If the force necessary for detaching the film is plotted against the peel length (distance), than the work of detachment is equivalent to the area under the force/distance curve (Figure 6.10).

The work of detachment for a given strip of coating with area A can be calculated by Equation 6.1:

Equation 6.1: $$G = \frac{F \cdot a}{A} = \frac{F}{B}(1 - \cos\Theta)$$

with

G = work of detachment in [N/mm]

a = Peel length

B = Width of the films

A = a·B = Area of the detached coating film

In case the coating film is pulled off under an angle of 90°, Equation 6.1 can be simplified to Equation 6.2.

Equation 6.2: $$G = \frac{F}{B}$$

The incorporation of a textile fabric into the coating involves external influences like interfaces between the textile and coating material. These external factors can be reduced by special imbedding techniques or by preparing a loose end of coating film without any fabric, so that it is pulled off the substrate in original state. In the latter case the work of detachment is composed in a complex manner. The work of detachment has to be completed by the deformation work of the coating, which includes work for straining and bending, as described by *Kinloch* (Equation 6.3) [14].

Equation 6.3: $$G = \frac{1}{B}\left(\frac{dU_{ext}}{d_a} - \frac{dU_s}{d_a} - \frac{dU_{dt}}{d_a} - \frac{dU_{db}}{d_a}\right)$$

G: Total peel energy

dU_{ext}: External work

dU_s: Stored strain energy of the peel stripe

dU_{dt}: Energy dissipated during tension stress exposure

dU_{db}: Energy dissipated during bending stress exposure close to the peel front

d_a: Peel length

B: Width of the peel off coating stripe

The work necessary to strain the coating can be obtained in separately performed tensile stress experiments on free films. It is reported that this energy portion can be subtracted from the total energy by using special software [15].

Instrumented knife cutting test (hesiometer)

This method represents an instrumented version of the popular knife cutting test. The knife has a carbon steel wedge blade, which is placed between coating and substrate. By motor-driven feed it removes the coating automatically (Figure 6.11).

The required force to detach the coating can be plotted as a function of the distance, allowing obtaining the total energy for removing the coating, similar to the peel test (Equation 6.4).

Equation 6.4: $$\frac{F \cdot s}{A} = \frac{F}{B}$$

with

A = B·s = Area of detached coating film

Further information can be received by fracture analysis (hard-brittle, tough-elastic). The mode of fracture and measured force values depend strongly on the film thickness, as could be seen from tests of wedge-shaped film thickness profiles. By comparing the breakage patterns with peel force values it could be shown that there is a link between the share of cohesive and adhesive failure and film thickness [16].

6.2.3.3 Causes and prevention of coating adhesion failure

The main causes for adhesion failure on plastic substrates, which have already been described in earlier chapters, are listed below:

- Substrate, component (see Chapter 2.3.3)
  - Ingredients (additives, release agents, molecular weight distribution, content of recycled material)
  - Processing parameters (injection pressure and speed, post pressure...)
  - Processing additives (external release agents...)
  - Storage, handling

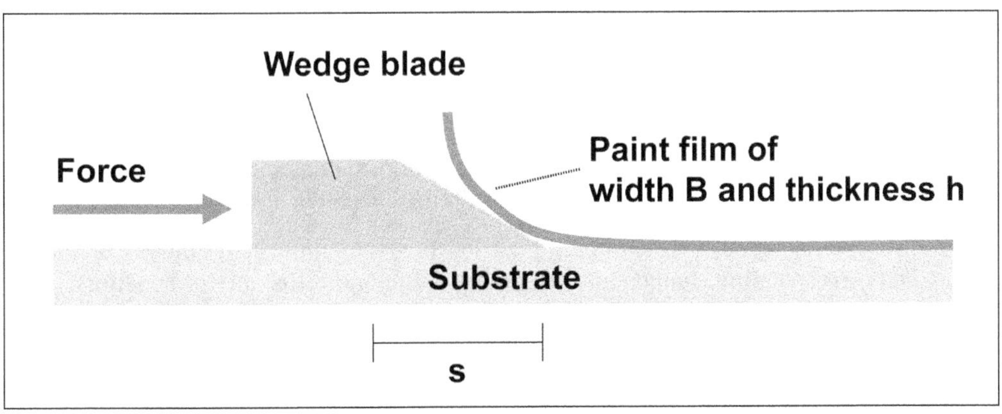

Figure 6.11: Scheme of the hesiometer test

- Pre-treatment (cleaning and activation, see Chapter 3)
  - Cleaning insufficient
  - Fingerprints
  - Power wash-contaminated
  - Inadequate activation process or processing parameters
- Coating material
  - Formulation/material unsuitable for the substrate
  - Handling, storage
- Coating application
  - Mixing ratio not adjusted properly
  - Film build too high/too low
  - Inadequate curing conditions (oven temperature, belt speed, climatic conditions)
  - Contaminations (delimitation of the paint shop)

6.3 Mechanical technological properties

During their operating life coating layers can be exposed to various stress factors causing deformation. This accounts especially for coatings on plastic substrates. The reasons for this are the overall higher ductility and lower strength of plastics compared to metallic substrates. In addition, many plastics are sensitive to solvents, expressed by either softening and/or embrittlement, often observed as stress cracking. The mechanical properties of the plastic substrate must not get worse after it has been coated. Of course a mechanical deformation of the coating must not cause any negative impact on its adhesion, functionality or aesthetic properties. Internal stress of the plastic substrate may also cause deformation, crack formation and affect adhesion of the coating, even in single layers of a multi layer paint body. Basically in these cases the moulding process has to be optimized, for example the tool temperature, and maybe also the exchange of the plastic for a material with lower CTE has to be considered (Chapter 2).

6.3.1 Impact and chip resistance

In contrast to coatings on metallic substrates, plastic coatings undergo non-destructive mechanical deformations not in a permanent but reversible way. Polymer substrate and coating are not deep drawn and bent permanently, but either exposed to vibration or fast impact. Painted plastic plates and components are examined in various stone chip and **impact tests**. Bending and deep drawing tests may give additional information and orientation about the flexibility of coated plastics, although mostly not being specified.

Automotive exterior surface areas being exposed to stone chip consist mostly of coated plastic. The corresponding components are mainly bumpers, grills, mirror housings, glazing and with some models fenders. Typically stone chip tests are executed as **multi impact test** according to ISO 20567-1 or individual car manufacturers specification. Similar as with automotive body coatings the mechanical performance of plastic coatings at deep temperature plays an important role, especially if substrates are used on which adhesion is hard to achieve. With these plastics often used, shock type stress at below-zero temperatures can easily result in chipping of a coating. The effect of coatings on plastics impact strength at various temperatures is investigated by the instrumented **high speed impact test**, according to ISO 6603-2. In this test a metal bolt is pushed with defined and instrumentally controlled speed through a coated plastic panel, and thereby the force is measured along the way through this panel. Plotting the force against the way up to failure results in a curve, of which the area underneath represents the work of break (Figure 6.12).

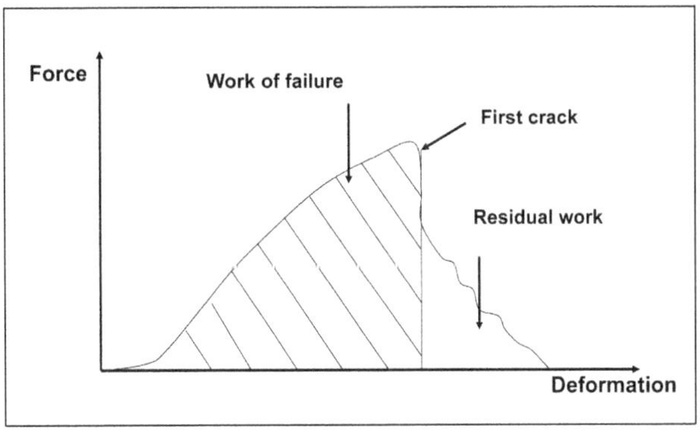

Figure 6.12: Instrumented high speed impact test according to ISO 6603-2

The damage formation can be recorded with a high speed camera, as shown in Figure 4.14 in Chapter 4.6.1.1.

Basically it is required that the performance of coated plastics is equal to the one in non-coated state.

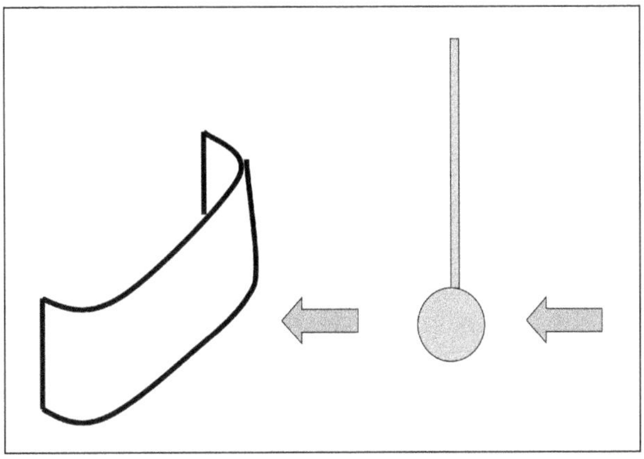

Figure 6.13: Scheme of a head impact test with a bumper covering

Another important aspect of this test is to determine the brittle/tough transition temperature, detectable by a rapid increase of the work load up to breaking. In general the aim is to achieve a low temperature value for the brittle/tough transition, in order to allow a shatter proof breaking even at below zero temperatures.

For the simulation of accidents involving pedestrians, whole painted components are fixed upon skids to expose them to a **head impact test** (Figure 6.13).

6.3.2 Hardness, abrasion and scratch resistance

The hardness of coated plastic components is proved in practice quite differently. Packaging coated plastic goods may cause marks by impressing packaging material, especially if the topcoat has not attained its full degree of cure. This **block resistance** can affect the stacking of freshly painted materials and is generally tested according to ISO 4622 or ISO 3678 [17]. Block resistance as a mechanical performance characteristic is presumably connected with coatings surface hardness, so it should be helpful to test the **pendulum damping** according to ISO 1522 or **micro-indentation hardness** method according to ISO 14577-1.

Another important performance characteristic influenced by hardness is the **polish ability** of topcoats and clear coats. Automotive body top coats, due to their higher hardness compared to plastic coatings, are easier to polish than topcoats for plastic parts. With respect to specified demands concerning low temperature impact strength, topcoats for plastic substrates cannot be adjusted to the same level of hardness. As a consequence clear-coats for plastics are softer and therefore can only be polished to a limited extent. Therefore hardness and flexibility of plastic coatings have to be adjusted carefully so that they are balanced out in a proper ratio.

The **abrasion resistance** of plastic coatings is important because of various abrasive stresses caused by dirt, sand, grit and manual handling. The latter is also relevant in the field of coated electronic consumer goods. But also coatings on aircrafts may suffer from abrasive processes as a consequence of their high speed, for example on edges of tail fins. which may be made from polymeric substrates.

It is assumed that abrasion of coatings is caused in a complex way by polymer mechanical factors. Obviously the structure of polymer molecules, including their cross-linking and resulting viscoelastic character, play a decisive role, as well as formulation parameters like the filler content, and external factors like the temperature. The investigation of the abrasion has to respect all these factors and is performed at best in a way that simulates real conditions. The **Taber-Abraser method** is described in ASTM D 4060 and DIN 53754. It uses two turning abrasive wheels removing coating material, depending on its composition and cure status, kind of wheel size, granulation and hardness of sand paper used, as well as load and test duration. The mass loss of coating material can be measured as a function of the number of cycles up to the degradation of the coating layer down to the subsurface.

Abrasive media that do not remove coating material to a significant extent do nevertheless represent a severe stress factor for coating, since they more or less always scratch the surface causing a reduction of gloss. The simulation of this kind of stress is usually performed in **scratch resistance tests**, which has been the focus of automotive clear coat producers for a long time. As painting of bumpers, mirror housings and other integrated components in body colour have become standard, clear coats for plastics have to fulfil increased demands for scratch resistance. The issue of scratch and abrasion resistance additionally gains in importance, as glazing (polymer glazing) of plastic substrates becomes more widely used. In order to improve scratch and weather resistance of polycarbonate glazing, thin layers of polysiloxanes are often applied [18]. Scratch resistance is a characteristic that has to be regarded in a differentiated way. For example, a distinction can be made between initial scratch resistance and scratch resistance after a certain service life, which may cover the effect of climate changes. Especially the scratch resistance after exposure to car wash is of great importance. Special attention should be paid to the time dependent change of scratch damage pictures by recovery of the coating). A useful survey of test methods is given by reference [1]. Basically these methods can be transferred to plastic coatings. Resistance to wash scratches can be examined by a car wash simulation procedure called **Amtec-Kistler test**. In this test coated panels are stressed by rotating brushes with polymer made bristles, which are sprayed with an aqueous slurry of quartz sand (1.5 g/l). The evaluation criterion is the retention of gloss, which for a scratch resistant coating of course should be as high as possible. It could be shown that a comparable simple laboratory test method as the **Crockmeter test**, even if performed with a plain hammer and abrasive fleece, can give valuable information concerning car wash scratch resistance [19]. On a more scientific level the scratch resistance can be investigated with the **Nanoscratch test,** in which an instrumented single scratch by a nano-indentor is applied to the coating, allowing to measure fundamental polymer physical properties simultaneously [20]. Imaging of the scratch process, in connection with an analysis of the physical data along the scratch way, allows a comparison of visible local damage details and physical properties. In general plastic coatings show a stronger tendency to reflow and self heal than automotive body coatings. Viscoelastic basic properties of clear coats for plastics have been correlated with their scratch resistance test results by *Ryntz* et al [21]. In this publication the most significant material influence parameters that could been identified as key properties for scratch resistance were the **essential work of fracture,** as obtained by tensile testing, and the **recovery** performance, as measured in the **Ford five-finger** test, a sort of five-fold single indenter method. Nano scratch test based investigations revealed the so called **critical depth to fracture** as the most engraving factor for the scratch resistance of clear coats on plastics. Thus indenters seem to penetrate deeper into clear coats on plastic substrates, compared with the less flexible automotive body clear coats, for which the critical load of fracture (related to the area of intention) is a more decisive factor [22].

The numerous performance test methods for painted plastic components could only be presented here in excerpts. Some useful further information can be found in the specifications of producers of consumer electronics and car makers, as well as in publications of known technical journals.

6.4 References

[1] H. J. Streitberger und K. F. Dössel (editors), Automotive paints and coatings, 2nd edition, Wiley-VCH Verlag, Weinheim, 2008, chapter 12

[2] S. Paul, Painting of Plastics: New challenges and possibilities, Surface Coatings International Part B: Coatings Transactions, Vol. 85, B2, 2002, p. 79

[3] J. A. Bagdachi, Journal of Coatings Technology, Vol. 69, No. 870, July 1997

[4] M. Rasche, JOT 9 2007, p. 46

[5] G. L. Nelson, K. N. Gray, Waterborne & Higher Solids Coatings Symposium, sponsored by Southern Society for Coatings Technology and University of Southern Mississippi, Feb. 5-7, 1986, New Orleans, LA, USA

[6] W. Brockmann, Adhäsion, 22 (1998) p. 6

[7] S. S. Voyutzki, V. L. Vakula, J. Appl. Polym.Sci., 7 (1963) p. 475

[8] S. S. Voyutzki, J. Adhes. 3 (1971) p. 69

[9] L. H. Sharpe, J. Adhesion 67 (1998) p. 277-289

[10] A. Goldschmidt, H. J. Streitberger, BASF Handbook Coating Technology, Vincentz, 2007, Chapter 3

[11] R. A. Ryntz, JCT CoatingsTech, 2, 18 (2005) p. 30

[12] G. Meichsner, Lackeigenschaften messen und steuern, Vincentz Network, Hannover, 2003, p. 135

[13] J. S. Meth, D. Sanderson, C. Mutchler, S. J. Bennison, J. Adhesion, 58 (1998) p. 117-142

[14] A. J. Kinloch, C. C. Lau, J. G. Williams, International Journal of Fracture 64 (1994) p. 45-70

[15] Adhesion and fracture of polymeric coatings, D. R. Moore, Surface Coatings International Part B: Coatings Transactions, Vol. 84, B4, 243-336, November 2001

[16] W. K. Asbeck, Journal of Coatings Technology Tech., 12 (2005) p.48

[17] T. Brock, M. Groteklaes, P. Mischke, European Coatings Handbook, 2nd edition, Vincentz, Hannover, p. 393

[18] K. Brammer, A. Ramlow, Kunststoffe, 8 (2006) p. 84

[19] M. Fies et al., Farbe und Lack, 118, 4/2008, p. 22

[20] E. Klinke, M. Kordisch, G. Kunz, Claus D. Eisenbach, Farbe und Lack, 108, (2002) p. 54-60

[21] R. A. Ryntz, D. Britz, Journal of Coatings Technology, 74, 925 (2002) p. 77

[22] V. Jardet, R. A. Ryntz, Journal of Coatings Technology Research, 2, 8 (2005) p. 591

7 Coating of plastics - how to detect and avoid failures

Coating failures may lead to disturbances of the whole production process, creating rework and downtime thereby increasing production costs and product pricing. This can also lead to company loss of image and regress claims of customers. For these reasons it is necessary to identify and analyze failures as early as possible, to work out and install appropriate counter measures in order to avoid failures in future [1].

Is it possible to achieve this high demand? In many cases coating failures can be identified, but despite all efforts one may not always achieve a correlation of root cause and effect. Failure analysis is limited due to the fact that samples are mostly not perfect. In addition to that, time and financial resources for this purpose are limited. Moreover is our understanding of coating technology still not complete, meaning that it is not possible in every case, why one system works well, when another might fail. Here it may be helpful to keep an open-minded, curious and investigative attitude. With a solid understanding of the whole manufacturing process for painted plastic components, as well as using modern powerful analytical techniques proper samples with effective communication between all partners along the process chain, it should be possible to overcome the above mentioned limitations and solve most of the problems arising with coating plastics [2].

The identification and avoidance of coating failures is a complex task, which requires an intense cooperation of all people involved in the production process of coated objects. A sustainable, successful failure analysis should not be limited to the application of analytical techniques, but is characterized by a strategically and systematic approach, which to a high extent can be learned [3].

At first it is important to describe the problem correctly, as perceived with the keen eye, without jumping to conclusions. An unprejudiced attitude will help the observer to describe the problem without mixing it with a too early diagnosis. Based on a neutral description many failures can be identified in the next step, by microscopy. Additional simulation experiments in the coating laboratory may often be helpful to encircle possible causes. Amplification of the assumed root cause may lead to a further clearing up. In other cases one must carefully collect jigsaw pieces before it is possible to recognize the whole picture. In order to achieve a sustainable elimination of the failure, it is necessary, to recognize the responsible material- and process deviation as possible causes. One of the preconditions for this is a profound documentation, executed by the employees at the plant (see checklist under Chapter 7.5). The corrective measures, that have been defined, often cannot be put into action right away. In order to minimize production loss and rejects, it may be of advantage to initiate first aid measures and carry out solid, sustainable measures later.

Wilke/Ortmeier: Coatings for Plastics

ISBN: 978-3-86630-859-6

In many cases coating failure analysis leads to clear root causes. If there are no other constraints and the identified origin is connected exclusively to negative consequences, then avoiding the failure is simply done by elimination of this root cause. This may not always be possible without side effects. But in general there are some basic prevention measures, which should help to reduce frequently occurring failures. Among them are maintenance, cleaning and control measures.

What are the most frequent failures with coating plastics, how are they to detect and which measures can be taken to avoid them?

7.1 Inclusions/contaminations

These are mostly small, irregular elevations in the coating film, which are caused by small amounts of external solid material of variable size, shape and size distribution.

These failures are either detectable initially, or their later effects on coatings performance for example against humidity, visible as blisters. Therefore the failure analysis of inclusions/contaminations cannot be separated properly from that of blister formation, and consequently they are treated here in as common.

7.1.1 Kinds of inclusions

The most occurring causes for inclusions are fibres and spray dust. But also metal particles, gases, hair, sanding dust as well as undefined dust and dirt particles are frequently found contents in microscopic analysis of this kind of failure locations. Coating inherent particles are another optional cause, giving rise for a lack in material homogeneity. Checking wet samples by casting on glass, as well as by chemical analysis of sieve residues (IR, EDX) may help to find indicators of inhomogeneity like gel particles or abrasion of grinding beads that may have been created within the coating production process. Also crystalline urea formed by hydrolysis of over-aged isocyanate catalyst can be detected or excluded in this way. In the following the individual kinds of inclusions and causes shall be treated:

Fibres

The presence of fibres as a cause of paint failures can have different origins. Plastic parts can be exposed to fibres at different places along the entire production process chain, beginning at the storage in raw state, to their cleaning and putting on skids. The air stream of pre-treatment, painting, flash-off and curing zones are potential risks of fibre contamination.

Figure 7.1: Light microscopic image of a fibre inclusion in cross sectional view Source: Lackfabrik Karl Wörwag GmbH & Co.KG

Even the storage of raw parts between layers of cardboard or weakly glued paper can result in a contamination with fibres. Fibres and hair can also be caused by unsuitable gloves, cleaning tools and cloths as well as clothing worn by the paint shop staff. Other potential sources of fibres are unsuitable cleaning tools in the surrounding of the paint line, for example to wipe floors and other surfaces in the facility.

Figure 7.1 shows a light microscopic image of a typical fibre inclusion. The fibre is well embedded in the coating, often barely visible in top view as a slight elevation. The type of fibre, as emitted from the clothes or gloves can be identified quite easily by scanning probe electron microscopy (SEM) (Figure 7.2 and Figure 7.3). Cotton fibres show a flat belt shaped structure with twisted locations. Fibres from synthetic fabrics can have different cross sections, whereas the structure is constant all over the whole fibre length.

Cotton fabrics by far emit more fibres than synthetic fabrics. For this reason the latter are preferred for coating applications than cotton materials. Gloves in principle should not have any open cut-off edges.

Spray dust

Unevenly distributed particle derived elevations in a topcoat or clear coat mostly can be referred to spray dust in the basecoat cabin or primer cabin (Figure 7.4 and Figure 7.5). Skids, conveyor belts, the air of ovens and cabins, contaminated with spray dust, are frequent sources of multicolour particles found in coatings. Of course there also are colourless spray dust particles, for example caused by grey primers

The load of spray dust caused by over spray in general increases, when the cabin temperature is too high, the spray viscosity is too low, and/or the solvent composition is balanced out improperly, as it flashes off too fast.

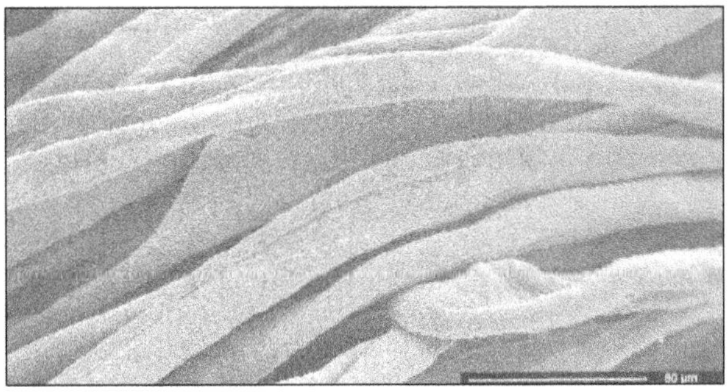

Figure 7.2: SEM of cotton fibres Source: BASF Coatings GmbH[4]

Figure 7.3: SEM of synthetic fibres Source: BASF Coatings GmbH[4]

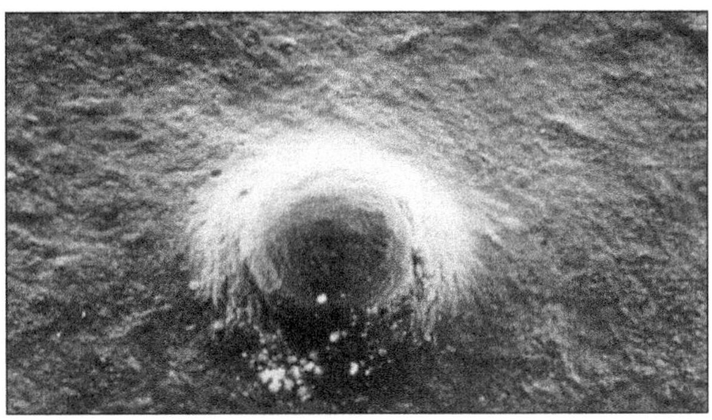

Figure 7.4: SEM of spray dust inclusion in top view, SEM-picture Source: BASF Coatings GmbH[5]

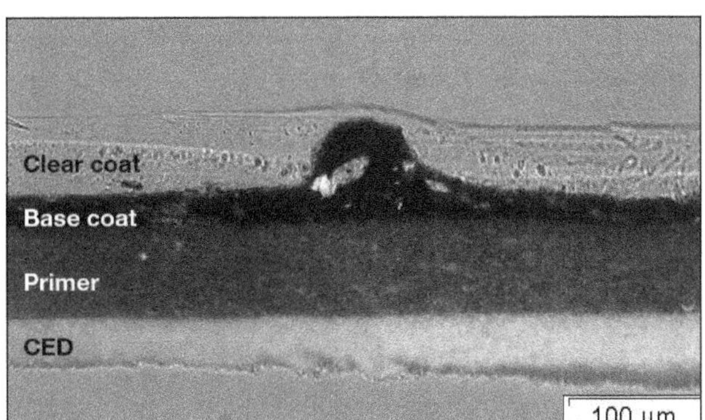

Figure 7.5: SEM of spray dust particle in cross sectional view, light microscopy picture Source: BASF Coatings GmbH[6]

Paint residues and decomposition products

Frequent sources of inclusions are paint deposition and dirt in spray pistols, circle lines and pressure vessels. Other potential causes are sanding dust, composed of paint or plastic material. It is also possible that the coating material is damaged within conveyance from the container to the atomization device. So it has been proved that flowing through a closed circulation pipeline can impose a too high shear stress on aluminium-pigments, resulting in deformation of the flaky shape and, later in spots in the topcoat. Generally the risk to contaminate the conveyance equipment increases with a low paint throughput as well as with the use of inappropriate closed circulation pipeline pumps.

Further possible causes for inclusion of paint material are deviations from the product application guidelines and, as a result, the formation of agglomerates and decomposition products. Waterborne basecoats after over-ageing can form in-homogeneities by resin coagulation or pigment agglomeration, which may be observed as film imperfections. One of the possible sources of failure with 2K PUR coatings is the use of a non-adequate batch of catalyst. Hydrolysis of isocyanate catalyst during aging may lead to precipitation of urea crystals at the container opening and carbon dioxide formation, perceivable as pressure increase in the container.

Undefined dirt particles

The source of dirt inclusions cannot always be detected. The following possible causes may be considered:

- Dust on the substrate surface (insufficient cleaning)
- Dusty external air, sucked in by negative pressure in the paint shop
- Defect fresh-air filters and insufficient separation of the painting zone from external areas increase the risk of a contamination with dusty air from the environment.

7.1.2 Inclusion analysis

Like craters and other surface defects, inclusions should not be touched with fingers and also not be grinded. To examine film defects, they are at first inspected with a stereomicroscope of magnification 40 to 80-times. The inclusions are prepared in thin slices and finally inspected with a microscope of higher magnification (Figure 7.6 and 7.7).

After fixation on a slide with adhesive tape they can be inspected with IR spectroscopy, light microscopy, SEM and X-ray microanalysis.

Prevention of inclusions

The following measures are recommended:

- A practice-oriented inspection of paint raw materials
- Cleanness in the paint supply facilities. Regular maintenance and functional testing of spray booths (including airflow, ionization, etc.)
- Regular exchange of filters
- Thorough cleaning and maintenance of atomization equipment, pumps and pipelines
- Cleaning cloths and devices, clothes, gloves made of suitable materials (synthetic fabrics, no cotton)

- Use all components in a well-tempered state
- In case of frequently occurring inclusions:
  - Immediate microscopic inspection and/or spectroscopic analysis
  - Documentation of defective parts, sieve residues, including chronological sequence (batches, maintenance etc.)
  - After analysis: consequent elimination of failure causes.

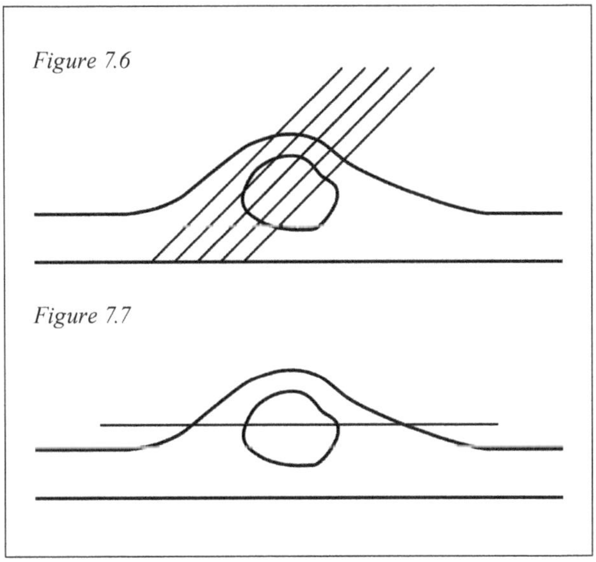

Figure 7.6 and 7.7: First cut and horizontal cut of a paint film inclusion

7.1.3 Blistering

Blisters filled with liquid normally form when salt residues on surfaces, maybe because of insufficient cleaning, give rise for osmosis. In a humid atmosphere, or after aging, this development may be caused by hand sweat, or by carry-over of cleaning agents from the last rinsing zone of a power-wash system. Blister formation may also be caused by connate water in troughs or storage in a too humid atmosphere. Another possible root cause for blister formation may be salt-like ingredients of the plastics bulk material. One example is the substance zinc stearate. Blisters on coated reinforced plastic substrates often contain nuclei from fibres and mineral fillers. In case of insufficient cleaning of the plastic components, often those blisters show crystalline salt residues on their backside, as can be demonstrated by SEM (Figure 7.8). An EDX analysis can then reveal the elementary composition of the salt residues, which can act as initiators for osmosis, causing weak humidity test performance.

Figure 7.8: Crystalline material in opened blister, SEM

Source: BASF Coatings GmbH[5]

Prevention of blister formation

In order to prevent blister formation, the plastic substrate surface should be dry and clean. The parts are stored at best under well tempered conditions. Pre-treatment has to start with well controlled cleaning operations and manual treatment of the components should only be allowed with suitable gloves (synthetic fabrics, no open edges). If cleaning is performed by power wash, the water in rinsing zones should frequently be controlled in terms of its electrical conductivity. If necessary, a higher frequency of exchanging rinsing water should be taken into account. Finally the power wash drying zone should be examined towards its efficiency.

7.1.4 Micro bubbles (pinholes, popping)

Micro bubbles in general are described as pinholes. They can be created by foam or other trapped gases. Sometimes the term pin-hole is generally used for popping. Solvents flashing off the paint film may cause micro bubbles, if the film gets closed before all solvent has left the coating, during the advanced phase of curing. Popping in a topcoat may take place, if solvents cannot leave the topcoat completely, due to an insufficient flash-off before the thermal curing starts. One of the possible causes for this is a too slow evaporation rate of the topcoats solvent composition. A too rapid viscosity increase of the wet film, caused by cure beginning to take place, may trap residual solvent as blisters in the curing film. Another potential risk for blisters is high film thickness, because all volatile components then have to pass a longer way of diffusion to get out of the film. Flash-off characteristics of a liquid paint is also influenced by the kind of paint application. In spray paints the solvent composition of the wet film depends on atomization parameters. For example the overall surface area of the spray dust droplets decreases with higher material flow rate and lower spray pressure. As a result slow evaporating solvents may not flash-off the spray dust fast enough and are enriched in the liquid film, giving rise for a higher extent of popping, compared to a dryer application caused by higher air pressure.

The same result may be obtained by changing the distance of atomization device and object surface, giving slow evaporating solvents less time to leave the spray dust. In all these cases the flash-off zone in a paint line often is not sufficient for a complete evaporation of the paints solvent. The outline and operation of spray cabins, drying and curing zone are other possible causes for the formation of pinholes and popping. Some factors interfering with solvents flash-off and inducing popping are:

- Temperature of the spray cabin too high
- Insufficient air circulation
- Rise of the oven temperature too steep

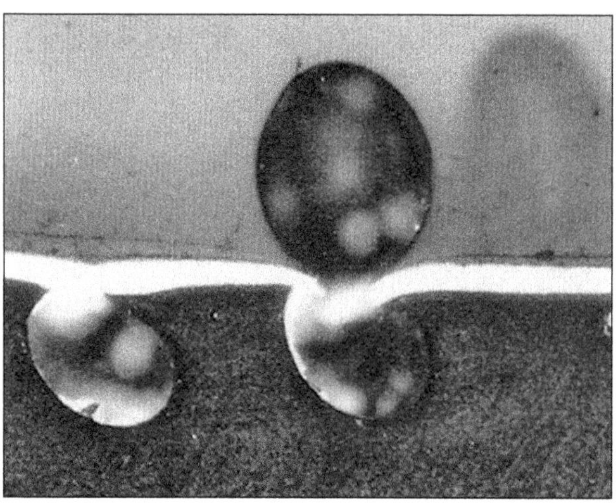

Figure 7.9: Bubbles in a substrate, penetrating into the coating layer Source: BASF Coatings GmbH[6]

Finally the origin of popping in a topcoat can also be located in a primer, which is not sufficiently dried before being over-painted. In analogy to the above treated case of a topcoat, popping can also take place in a primer, if the flash-off time is not sufficient to remove all solvent completely. Temperature dependent venting of bubbles, voids and pores in the substrate material may also give rise for pinholes and other failures in the primer. Some substrates contain volatile ingredients like residual monomers or other process chemicals (SMC,

RRIM etc., see Figure 7.9). Humidity content of substrates as polyamide and polyamide blends (PPO+PA) has to be balanced out before painting, in order to prevent pinholes. A potential relief of that problem is the conditioning of components made of these polymers.

Micro bubbles, prevention and first aid measures

- Identify failures by microscopy
- Inspect all coating and drying parameters, in particular film thickness (by microscopy)
- Check and control the paint temperature upon diluting
- Check the functionality of manometers, thermometers etc.

7.1.5 Wetting failures

Closely confined, mostly circular shaped wetting failures are usually called craters. The main physical cause of craters is probably a gradient of surface tension between failure zone and surrounding area. These can be caused by solid or liquid contaminations or by incompatibilities in the paint material (Figure 7.10). Contaminations sometimes can be identified chemically by instrumental analytical methods like TOF-SIMS or X-ray microanalysis. Extreme large craters are also called fish eyes.

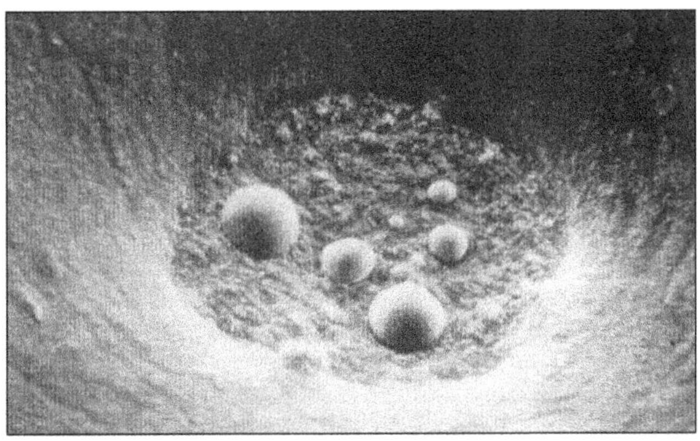

Figure 7.10: Crater with a contamination, SEM

Source: BASF Coatings GmbH[5]

Possible causes for craters in a primer are:

- Grease residues or other fat-like contaminations (silicones, fluorinated hydrocarbons) on the substrate surface
- Pores in the substrate
- Fibres sticking out the substrate

Figure 7.11: Light microscopic image of a wetting failure, caused by a long chain hydrocarbon

Source: Lackfabrik Karl Wörwag GmbH & Co.KG

Insufficient levelling of the primer may also promote the formation of craters, in addition to other factors. Unsuitable application of metallic basecoats can result in non parallel orientation of aluminium platelets sticking out of the basecoat layer. As a possible consequence the clear coat can be affected in its ability to wet the basecoat layer. Craters in a topcoat can be induced by bad levelling, a subcritical film thick-

Figure 7.12: Covering of surface defects

Source: BASF Coatings GmbH[4]

ness and by spray dust which is embedded in the primer layer.

Apart from craters and fish eyes there are wetting failures, which neither have a round shaped limit, nor are they caused by in-homogeneities of the paint material or by contaminations. Every coating has a certain limit of film thickness, under which the surface is not wetted anymore. If the film thickness falls below that limit during the painting process, then the subsurface is either not completely or unsuitably coated. This wetting limit belongs to the application quality criteria, especially of clear coats.

In addition to coating derived influences, plastic surfaces can have an inhomogeneous distribution of local wet ability, giving rise for local wetting failures (Figure 7.11).

Analysis of craters

The first step in an effective approach to analysis of craters is a suitable treatment of the defect. The surface to be investigated must not be touched nor marked with a pen. Prior to transportation the defect has to be covered ideally with a clean piece of paper with edges taped on the surface (Figure 7.12).

Identification of possible crater inducing substances, like silicon oils and fluorine-hydrocarbons, a powerful analytical tool has proven to be time of flight secondary ion mass spectroscopy (TOF-SIMS).

7.2 Disturbed interactions in the paint build up system

7.2.1 Coating adhesion failure

Adhesion failures of coatings on plastic substrates can partly be related to the kind of processing, by which components are manufactured, that have to be coated. For example it is possible that the injection moulding parameters are not suitable for coating purposes. See also Chapter 2.3.3.2 Recommendations.

Causes and prevention of coating adhesion failure

In the following the most of already above mentioned possible causes for adhesion failure of coatings on plastic substrates are listed briefly:

- Substrate, component (see Chapter 2.3.3)
  - Ingredients (additives, release agents, distribution of molecular mass, ratio of recycled material)

  - Processing parameters (injection speed, post injection pressure etc.)
  - External de-moulding agents
  - Storage, handling
- Pre-treatment (cleaning and activation, see Chapter 3)
  - Cleaning insufficient
  - Fingerprints
  - Power wash-zone contaminated
  - Activation or activation parameters insufficient
- Paint material
  - Formulation/material unsuitable for the kind of substrate
  - Handling, storage
- Coating application
  - Mixing ratio not adjusted properly
  - Film build too high/low
  - Curing conditions inappropriate(oven temperature, belt speed, climatic conditions)
  - Contaminations (delimitation of the coating line)

7.2.2 Crack formation

Cracks of different length and depth can result from defects in building up the paint layer body. If for example a flexible plastic substrate is coated with a rigid paint system, then crack formation can easily take place even under comparable low mechanical loads. The cracks will then be initiated in the coating and proceed in direction of the substrate. A crack that has been formed from the outer sphere of a coating is assumed to act as a wedge pushing towards the deeper zones of the coating and finally to attack the substrate itself.

Coating a plastic substrate with a paint material containing a too aggressive solvent composition can result in the formation of stress cracks. Some plastics like PS- and PC-blends as well as PMMA are prone to form stress cracks when brought into contact with esters, aromatics and ketones (see also Chapter 2, Table 2.10). Internal stress is another influence factor of crack formation. Depending on the cooling rate of the polymer melt in the final phase of injection moulding, the moulded plastic part shows a higher or lower level of residual stress. For this reason, a sufficient long cooling time is necessary in order to bring the polymer material to a state of thermodynamic equilibrium, which means a decrease of residual stress. Although in general such a low stress inducing injection moulding procedure is strived for, this is only partly feasible, as for cost reasons it is important to keep cycle times as short as possible.

Finally cracks in a coating can also take place, when the surface of the substrate has already ruptured. In this case the coating cannot cover the crack in the substrate.

7.2.3 Wrinkling

Wrinkling of a topcoat from the subsurface is an optical striking, not tolerable failure, affecting also the functioning of the layer composite (Figure 7.13).

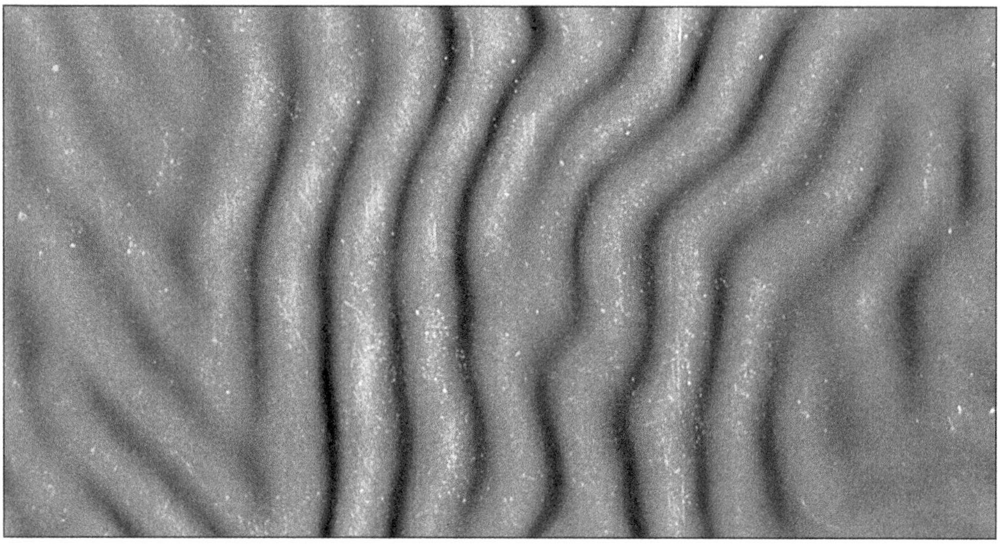

Figure 7.13: Wrinkling *Source: BASF Coatings GmbH*[5]

Wrinkling is mostly caused by dissolving of the layers (often both basecoat and clear coat) being removed from the subsurface as a result of

- Too aggressive solvent composition
- Insufficient flash-off of the clear coat
- Too high film build of the basecoat

As a consequence wrinkle build up can mostly by prevented by elimination of the above listed deficiencies.

7.3 Levelling and appearance defects

7.3.1 Orange peel effect

A coating with poor levelling and appearance can in extreme cases present itself as orange peel surface.

Possible causes that have to be considered are mostly located in the topcoat and sometimes in the primer. These are a too high viscosity of the coating or a too rapid flash-off due to fast boiling solvents. Levelling of a coating is mainly controlled by rheology of the freshly applied wet film. So there are some application and paint shop parameters to respect, which have a major impact on wet film viscosity. These are:

- Atomization parameters
- Substrate temperature
- Temperature of the application cabin, flash-off zone and dryer
- Film thickness

When working with 2K coatings material, pot-life can also be a contributor to poor levelling. Furthermore the generation of an orange peel like surface can also be promoted by factors of the plastic surface, like pores and a grain structure.

7.3.2 *Sagging*

Sagging of coatings is observed and at best evaluated at holes and edges of vertically painted surfaces. There are some factors that either alone or in combination with each other can be considered as responsible for the primary cause, a too low viscosity of the liquid coating film:

- Solvent composition flashing off too slow or
- Content of solvent too high
- Film thickness too high
- Flash-off time too short
- Atomization too coarse caused by the following possible issues:
  - Distance of atomization device to the surface too small
  - Spray nozzle too wide
  - Flow rate too high
  - Atomization air pressure too low

7.3.3 *Substrate marks*

Visible gates, flow fronts and weld lines on plastic parts can often not be covered completely with a coating. The paint raw component can also show telegraphed surface patterns like that of "tiger skin". In this sense, substrate marks should not be regarded as coating failures but as substrate surface failures resulting from deficient moulding processes. The influence factors of injection moulding affecting coating quality are described in Chapter 2.3.3. Substrate marks sometimes can be reduced over the coating formulation, for example by the kind and content of fillers, but mostly not removed completely. On the contrary, it may also happen that substrate derived failures are amplified by a coating, especially when using unsuitable solvents. Which solvents are most appropriate in this context, hast to be evaluated in any case carefully, depending on the substrate and status of the moulded part. Primarily these are solvents with solubility parameters different from those of the substrate and therefore only with little aggressivity towards the plastics surface (see Tables 2.5 to 2.7).

7.4 *Deviations of colour-, effect- and gloss*

Optical failures are gaining importance for coatings on plastics, due to an increase in aesthetic demands on surfaces, and the trend towards individualization and the added value of decorative coatings. This applies especially on components in body colour that have to be integrated into automotive vehicles. Deviations in colour may then be related to both colour samples and painted metal body. A lot of progress has been achieved over the last years concerning the alignment of colour standard panels and the agreement on colorimetric tolerances among OEM, component and paint suppliers. Although in some cases it was possible to attain a high level of control, and the situation of colour matching between plastic part and car body has been generally improved, in daily production colour deviations still occur and seem hard to be turned off completely. The colour coordinates and tolerable deviations have to be well defined. The colour specific application parameters of the coating suppliers laboratory produced products have to be matched to the application process of the customer. Deviations of paint shop parameters should be communicated to the coating supplier as soon as possible. But even with nearly perfect communication there may be still a certain potential of shift in those parameters.

This includes:

- Fluctuations in film thickness (in particular with effect colours and poor hiding straight shades)
- Spraying too wet/dry (wet: flow rate too high, atomization pressure too low, cabin temperature to low)
- Wet/dry spraying effect on colour
  - too dry: brighter, colour silvery, colour greyer
  - too wet: colour darker, higher intensity

Mottling is another coating failure, which also is strongly influenced by the paint application parameters. The following causes can be found frequently:

- Application of the basecoat too wet
- Film thickness of the basecoat too low
- Flash-off of the basecoat insufficient and/or
- Application of the clear coat too wet or solvent composition too slow evaporating.

A poor gloss of the clear coat layer can be caused by:

- Film thickness of the basecoat too high
- Viscosity of the clear coat too low
- Film thickness of the clear coat too low
- Basecoat not dry enough

7.5 Checklist for recording complaints

This checklist with the most frequent issues and two tables should be regarded as suggestion and may serve as a template to record coating failures, following reference [3]:

Checklist

Failure recorded on (date): ______

Customer: ______

Plant/paint shop: ______

Production program affected: ______

1. Subject of complaint?
2. Description of the observation (found out in a plant visit or by telephone/e-mail etc.?):
   - Qualitative/quantitative
   - What articles/components are affected?
   - On which part of components surface is the failure located?
   - Since when and for how long have the problems been observed?
   - Which paint materials and batches are involved?
   - Where there similar problems earlier?
   - How where samples obtained (important in dirt analysis)?
   - Number of painted parts?
   - Amount of applied paint?
3. Root cause analysis/deviation analysis

Table 7.1: Fault analysis - material

Material	Type/grade/ batch	Within specification – yes/no	Kind of deviation	Sample to bring – yes/no
Substrate, raw part (including polymer quality, release agents etc.)				
Detergents used in cleaning process/agents				
Adhesion promoter/primer (including catalyst, thinner, mixing ratio, flash-off, curing, film thickness)				
Basecoat (including thinner flash-off, film thickness)				
Clear coat (including catalyst, thinner, mixing ratio, flash-off, curing, film thickness)				
Topcoat (including catalyst, thinner, mixing ratio, flash-off, curing, film thickness)				

Table 7.2: Fault analysis – coating process

Plastic part	Within specification – yes/no?	Kind of deviation?
Injection moulding, cleaning, activation, ionization		
Primer application		
Basecoat application		
Clear coat application		
Topcoat application		
Others (for example air circulation, maintenance work, part handling and storage, production line stops, status and delimitation of the paint shop, activities in the environment of the paint shop (construction work, transport activities, cleaning works etc.)		

7.6 References

[1] B. Mayer, JOT 6 (2001) 32

[2] D. G. Weldon, Failure Analysis of Paints and Coatings, John Wiley & Sons, LTD, Chichester, New York, Weinheim, Brisbane, Singapore, Toronto, (2001)

[3] H.-J. Oslowski, Lehrgang Kunststoffackierung, Technische Akademie Esslingen, 26-27.05.1997

[4] Fehlstellen in der Automobillackierung und deren Ursachen, Pocket guide, BASF Coatings GmbH, (2002)

[5] Glasurit-Info, Der Ratgeber bei Lack- und Lackierschäden, Firmendruckschrift der Glasurit GmbH (BASF Coatings GmbH)

Authors

Prof. Dr. Guido Wilke, born in 1962, studied chemistry at the universities of Düsseldorf and Münster/Germany. After receiving his PhD in organic chemistry he spent one year as a post-doc at the University of Washington/Seattle, USA. In 1992 he joined BASF Coatings GmbH in Münster/Germany to work on the development of resins, automotive and plastic coatings. Since 2002 he is Professor for Polymer Materials at the University of Applied Sciences in Esslingen/Germany.

Dipl.-Ing. (FH) Jürgen Ortmeier, is currently Global Research Director Karl Wörwag Lack-und Farbenfabrik GmbH & Co. KG, supplier of automotive and non-automotive coatings. In this capacity he directs strategic research and development activities for the organization and its global subsidiaries. His first contact with industrial plastic coatings accompanied his chemical engineering studies at Mercedes in Sindelfingen, Germany. After graduation to a coatings engineer in 1984, his career path leads him from the application to the development of specialized plastic coatings during his 25 year tenure with Wörwag GmbH. The experiences shared in this book are the result of 25 years of comprehensive knowledge of the automotive and non-automotive industry, and its related processes including plastic and metal pre-treatment, powder and liquid technologies.

U

V

W

www.ingramcontent.com/pod-product-compliance
Lightning Source LLC
LaVergne TN
LVHW081330021026
849161LV00008BA/1830
* 9 7 8 3 8 6 6 3 0 8 5 9 6 *